Forecasting Volcanic Eruptions

Edited by Angelo Paone and Sung-Hyo Yun

Published in London, United Kingdom

IntechOpen

Supporting open minds since 2005

Forecasting Volcanic Eruptions
http://dx.doi.org/10.5772/intechopen.77483
Edited by Angelo Paone and Sung-Hyo Yun

Contributors
Uriah Lar, Mohammed Tsalha, Isah Lekmang, Cedric Longpia, Prantik Mandal, Sang Koo Hwang, Leonel Vega, Angelo Paone

First published in London, United Kingdom, 2020 by IntechOpen
IntechOpen is the global imprint of INTECHOPEN LIMITED, registered in England and Wales, registration number: 11086078, 7th floor, 10 Lower Thames Street, London, EC3R 6AF, United Kingdom
Printed in Croatia

British Library Cataloguing-in-Publication Data
A catalogue record for this book is available from the British Library

Additional hard and PDF copies can be obtained from orders@intechopen.com

Forecasting Volcanic Eruptions
Edited by Angelo Paone and Sung-Hyo Yun
p. cm.
Print ISBN 978-1-78984-029-2
Online ISBN 978-1-78984-030-8
eBook (PDF) ISBN 978-1-78984-710-9

For EU product safety concerns:
IN TECH d.o.o., Prolaz Marije Krucifikse Kozulić 3, 51000 Rijeka, Croatia, info@intechopen.com or visit our website at intechopen.com.

Meet the editors

Angelo Paone received his BS and MS at the University of Naples Federico II, his English Advanced Certificate at Kingsway College in London, specialization course abroad (USGS, VA, Reston), and his PhD at USGS and at the University of Naples Federico II. He did a short postdoc at the American Museum of Natural History of New York, a Marie Curie Fellowship (individual postdoc) at Bristol University, and an Italian postdoc at the University of Naples Federico II. Since 2008 he has been a teacher in science at the Italian College Liceo Ettore Majorana Pozzuoli and a research professor at Pusan National University.

Dr. Sung-Hyo Yun is a professor at the Department of Earth Science Education, Pusan National University (PNU). He obtained his Bachelor of Science diploma in 1980 at the College of Education, PNU, and his Master of Education diploma in 1982 from the Graduate School, PNU. Dr. Yun received his PhD in Geology and Volcanology 1987 from the Graduate School, PNU. Dr. Yun was president of the Petrological Society of Korea and the Korean Society of Volcanic Hazards Mitigation.

Contents

Preface

This book discusses the behavior of several volcanoes worldwide. It refers to those volcanoes that are quiescent and/or active and that can transform the lives of those that live around them when they erupt. The development and application of new approaches and technologies will provide new visions about how volcanoes work so that a novel volcanic model can be built.

The authors who collaborated on this project have summarized their experience and present advances related to the assessment of hazards and risks associated with volcanic parameters. The book contains five chapters that cover important research aspects of hazardous volcanology. The first chapter is an introductory chapter prepared by the editor and looks at the present state of the art of volcanology with suggestions for the future of volcanic structures and the ability to forecast volcanic eruptions.

The next chapter presents the Jos and Biu Plateaux volcanic provinces occupying the northeastern half of Nigeria, bordering the Cameroon Volcanic Line, which has a conspicuously visible number of dormant volcanoes with no reported activity. The several incidences of volcanic eruptions along and close to the Cameroon Volcanic Line are pointers to the possible reactivation of any of the dormant volcanoes in Nigeria.

Chapter 3 presents experience of Deccan volcanism at 65 Ma when it moved over the Réunion hotspot, which altered lithospheric structures below the Kachchh rift zone. To quantify the influence of Deccan volcanism on the crust-mantle, this kind of study is very important to evaluate the difference between the components participating in the formation of such magmas.

Sang Koo Hwang illustrates the evolution of the Guamsan caldera associated with the Guamsan Tuff and rhyolitic intrusions in Chapter 4 and explains the evolution and formation of such caldera.

In Chapter 5, Leonel Vega, on the other hand, puts forward a new methodology for the case of volcanic risk computation. Most of the scientific, technical, and economic effort has been oriented towards the evaluation of threats, with a number of methodological considerations for assessing vulnerability and much less risk, so that integral volcanic risk can be computed.

The final chapter by the editor Angelo Paone studies the volcanic evolution, hazards, and risks of the Somma-Vesuvius volcano.

The editor expresses his thanks to all the participants for their valuable contributions and to Ms. Nina Kalinic Babic for her assistance in finalizing the book.

Acknowledgment also goes to IntechOpen staff members for their help in completing this book and other publications for free access.

Angelo Paone and Sung-Hyo Yun
Pusan National University,
South Korea

Chapter 1

Introductory Chapter: New Advances and Challenges

Angelo Paone and Sung-Hyo Yun

A volcano is a peculiar system where the evolution must be determined. Since a volcano is a complex and chaotic system, any volcanological information have to be taken in fully consideration in a broad volcanic model. Such task is not easy and many active and quiescent volcanoes must be watched attendally, and each volcanological parameter has to be studied and evaluate its influence on the broad hypothetically volcanic model at the unisono. This method could be a difficult endeavor, but it is worth to try.

1. Forecasting volcanic eruption

A major issue in the study of active volcanoes is the link between eruption history, the changes in magma compositions and volumes observed on the surface, and the nature and the time scale of magma differentiation processes underground [1, 2]. It is the latter which determine the timing and the style of eruptions, and which are therefore fundamental to models of how particular volcanic systems behave and evolve, and hence determine hazard mitigation. The immediate targets are those volcanoes that are active, have a history of devastating eruptions, and are in highly populated areas. To make best use of the latest analytical technology, it is also important to work on a well-studied volcano with an established stratigraphy and rock composition. Studies of volcanic samples, experimental investigations, and theoretical modeling are providing insights into the dynamics of magmatic systems and how the plumbing system evolves with time, giving a physical framework with which to interpret volcanic phenomena. All the volcanic processes evolving before and during an eruption lead to the variation of geochemical and geophysical parameters [3]. Hence, computational and analytical facilities, instrumentation, and collection of comprehensive observational, geophysical, geochemical, and petrological datasets associated with recent volcanic activity have enormously improved our view on how volcanoes work [4]. Many active volcanoes have been studied in terms of volcanic forecasting using only some of these techniques. The challenge is to try to employ a volcanological methodology where the main chemical, petrologic, physical, and geophysical parameters are linked all together in order to build a framework where the history of magma ascent velocities, the time scales of magma differentiation, the past and present situation of the magmatic conduit, and the present state of degassing must be considered simultaneously. Such conceptual scenario can be commensurate with monitoring quiescent volcanoes, thus forecasting volcanic eruptions. The papers presented in this International Volcanological Special Issue consider the characteristic features of a single volcano and/or a province of volcanoes on earth, in terms of a future volcanic activity. The technical methods used are wide and innovative as well as traditional concerning the knowledge presented for each paper and therefore worth to study. In this book by the title "Forecasting Volcanic Eruptions," we collect some example of less known volcanic systems and other well-known volcanic system such as Somma-Vesuvius volcano. Lar et al. [5] study the Jos and Biu Plateaux volcanic provinces occupying the northeastern half of Nigeria, bordering the Cameroon volcanic line, scattered with conspicuously visible number of dormant volcanoes with no reported activity. The several incidences of volcanic eruptions along the close-by Cameroon volcanic line are pointers to the possibility for the reactivation of any of the dormant volcanoes in Nigeria. Mandal [6], on the contrary, acquaints with the Indian plate that experienced the Deccan volcanism at 65 Ma when it moved over the Réunion hotspot, which has altered lithospheric structure below the Kachchh rift zone (KRZ). To quantify the influence of Deccan volcanism on the crust–mantle, Mandal [6] focuses on the delineation of the upper mantle structure below the KRZ, through the modeling of crust-corrected P-residuals and P-wave teleseismic tomography. The reduction in seismic velocity in the upper mantle could be explained by the presence of trapped carbonatite/partial melts related to the Deccan volcanism. The influx of volatile CO_2 emanating from the carbonatite melts in the asthenosphere might be generating lower crustal earthquakes occurring in the KRZ. Furthermore, Hwang [7] illustrates the evolution of the Guamsan caldera associated with the Guamsan Tuff and rhyolitic intrusions. The Guamsan Tuff consists of dominant ash-flow tuffs with some volcanic breccias and fallout tuffs. The breccias comprise block and ash-flow breccia near a vent and caldera-collapse breccia near a ring fracture. The lower member of the ash-flow tuffs is produced from pyroclastic flow-forming eruptions with any ash-cloud falls on the flow units, whereas the upper member is formed by many ash-flows from boiling-over eruptions. The rhyolitic intrusions are divided into intracaldera plug and ring dikes. The volcanic activities in the caldera exhibit the volcanic processes along a caldera cycle together with eruption types during 63.77–60.1 Ma. The activities began with pelean eruption that occurred with block and ash-flows from lava dome collapse, progressed through expanded pyroclastic flows and ash-cloud falls by pyroclastic flow-forming eruptions from a single central vent, and transmitted with non-expanded ash-flows from boiling-over eruptions along multiple ring fissure vents. Then the caldera collapse induced any translations into multiple ring fissure vents from an earlier single central vent. The boiling-over eruptions were followed by effusive eruptions along which rhyolitic magma was injected as a small plug and ring dikes with some lava domes on the surface [8–10].

On the other hand, Vega [11] puts forward a new methodology in the case of volcanic risk computation. Most of the scientific-technical and economic efforts have been oriented mainly toward the evaluation of threats, with few methodological considerations for assessing vulnerability and much less risk. In other cases, the threat and vulnerability are evaluated independently, which logically presents many difficulties for the integral risk assessment. It is also easy to verify that many of the studies called "vulnerability assessments" are only physical and functional characterizations and diagnoses of vital infrastructure and population. These characterizations can hardly be interpreted in terms of georeferenced indexes and/or maps of vulnerability that represent the spatial and temporal exposure of the elements exposed to each threat.

These were the questions and the academic challenge that led the PIGA Group of research in politics, information, and environmental management of the Universidad Nacional de Colombia, under the direction of the undersigned and with the institutional support of the Ministry of Environment of Colombia and the Autonomous Regional Corporation of Tolima, to investigate and study the threats, vulnerabilities, and risks faced by the population, the constructed elements, and the ecosystems of the area of influence of the Cerro Machín volcano, located in the Department of Tolima, Colombia.

Consequently, a new equation for the determination of volcanic risk is generated and adjusted, based on the valuation of intrinsic threat indexes according to their degree of intensity, duration, extension, and accumulation and of vulnerability indexes according to the degree of spatial and temporary exposure of the exposed elements (social, economic, institutional, and ecosystemic) and of their capacity of intrinsic and extrinsic response to the threats.

With these equations applied at the level of each pixel, and through the use of geographic information systems (GIS), the risk of damage is modeled and geospatially integrated for two analysis scenarios (start of crisis and eruption) of a possible eruption of the volcano Cerro Machín, which allows obtaining, according to the logical framework of the pre-established evaluation, both the total risk on each element exposed and the total risk generated by each threat.

As a final point, for each analysis scenario considered, the total risk maps as well as escape route maps and possible shelters for transient and/or final relocation of population and population centers are generated.

2. Volcanology at the centre of the people

Finally, Paone [12] shows how Somma-Vesuvius works with an outline on the state of the volcano. Somma-Vesuvius is a quiescent stratovolcano with a probability of Plinian-style volcanic reactivation. Its stratigraphy is well known in the last 40 ka BP. The volcanic products that are part of the Somma caldera are poorly studied. Conversely, younger products have been deeply studied together with the AD 79 Plinian eruption. The impact of a Plinian eruption has been studied and summarized. A simplified scheme is presented from what we can understand the volcanic hazard and risk that the volcano poses to the greater Neapolitan population. In the last 40 years, the demography around the Somma-Vesuvius volcano has increased; consequently, the volcanic risk has increased. It would seem that the Italian Civil Protection (ICP) has not influenced the population and the Italian authority with their massive work around Somma-Vesuvius (red zone). People still continue to build houses. Nowadays, the Somma-Vesuvius volcano does not seem to threaten people, or the people that live around Vesuvius are not afraid of the volcano. But as it is usual just in this moment that the work done and to be done must be speared to all Neapolitan people, working in the school to reach the family. People around Somma-Vesuvius tend to neglect the volcanic risk appearing around Vesuvius. So ICP, all must be much more attend about the behaviour of this Hazardous volcano.

Finally, any active or quiescent volcano must be studied by volcanologist very carefully and all the information must be passed to the people living around it in order to have the people at the centre of the Volcanology.

Author details

Angelo Paone* and Sung-Hyo Yun
Pusan National University, Busan, Republic of Korea

*Address all correspondence to: angelo.paone1@gmail.com

References

[1] Hawkesworth C, Blake S, Evans P, Hughes R, Macdonald R, Thomas L, et al. The time scales of crystal fractionation in magma chambers—Integrating physical, isotopic and geochemical perspectives. Journal of Petrology. 2000;**41**:991-1006

[2] Hawkesworth C, George R, Turner S, Zellmer G. Time scales of magmatic processes. Earth and Planetary Science Letters. 2004;**218**:1-16

[3] Sparks RSJ. Forecasting volcanic eruptions. Earth and Planetary Science Letters. 2003;**210**:1-15

[4] Cashman KV, Sparks RSJ. How volcanoes work: A 25 year perspective. Geological Society of America. 2013. DOI: 10.1130/B30720.1

[5] Lar U, Lekmang I, Longpia C, Tsalha M. Is a Volcanic Eruption Possible in Nigeria? 2019. This Volume

[6] Mandal P. P-Wave Teleseismic Tomography: Evidence of Imprints of Deccan Mantle Plume below the Kachchh Rift Zone, Gujarat, India. 2019. This Volume

[7] Hwang SK. Eruption Types and Processes in the Guamsan Caldera, Korea. 2019. This Volume

[8] Branney MJ, Kokelaar P. Pyroclastic density currents and the sedimentation of ignimbrites. Geological Society, London, Memoirs. 2002;**27**:152

[9] Sparks RSJ, Bursik MI, Carey SM, Gilbert JS, Glaze LS, Sigurdsson H, et al. Volcanic Plumes. Chichester: Wiley; 1997

[10] Fagents SA, Gregg TKP, Lopes RMC. Modeling Volcanic Processes. The Physics and Mathematics of Volcanism. Cambridge University Press; 2012. p. 448

[11] Vega L. Towards a New Conceptual and Methodological Approach for the Integral Evaluation of Volcanic Risk. 2019. This Volume

[12] Paone A. The Somma-Vesuvius Activity with a Focus to the AD 79 Eruption: Hazard and Risk. 2019. This Volume

Chapter 2

Is a Volcanic Eruption Possible in Nigeria?

Uriah Lar, Isah Lekmang, Cedric Longpia and Mohammed Tsalha

Abstract

The Jos and Biu Plateaux volcanic provinces occupy the northeastern half of Nigeria bordering the Cameroon Volcanic Line, dotted with conspicuously visible number of dormant volcanoes with no reported activity. These dormant volcanoes represent potential future eruption sites. The ejecta materials of these volcanoes are essentially basaltic in composition and consist of sequence of pyroclastic materials, basalts, scoria and ash and are formed by strombolian and effusive styles of eruption. The volcanoes are represented by well-preserved cones and lava flows. In places the lava flows have been lateritized and eroded leaving remnants of weathered basalt boulders and a number of plugs and dome-like outcrops lacking any preserved cones. The basalts display essentially similar compositions consisting of phenocrysts of both olivine, plagioclase (bytownite–labradorite), with minor pyroxene (diopside-augite) embedded in a groundmass of plagioclase laths (labradorite), and accessory magnetite, ilmenite, k-feldspars, and volcanic glass. Geochemical data shows that these basalts are mainly alkaline olivine basalts derived from the deep mantle source enriched in incompatible elements similar to that of the Ocean Island basalts (OIB). Preliminary ^{40}Ar-^{39}Ar ages on the some of the basalts revealed Quaternary ages (Pleistocene epoch). The significant change in the composition of the Pidong Lake marked by decreasing pH is indicative of a probable input of juvenile fluids into the Lake. Also, the several incidences of volcanic eruptions along the close-by Cameroon volcanic line are pointers to the possibility for the reactivation of any of the dormant volcanoes in Nigeria. This work focuses on the need to assess the hazard level of some of these volcanoes for effective monitoring, disaster preparedness and land use planning as more people live and farm in these potentially endangered volcanic prone areas, unaware of the inherent risk.

Keywords: volcanoes, eruption, Jos Plateau, Biu Plateau, Nigeria

1. Introduction

An inventory of all the volcanic prone areas on the Jos and Biu Plateaux were carried out, followed by detailed update of the Geology of two major volcanic areas on the Jos Plateau volcanic provinces (Kassa and Kerang volcanoes). The geochemistry (major, trace and REEs compositions) of these volcanoes were determined and a few dating using ^{40}Ar-^{39}Ar technique were performed on the Kassa basalts. Hydrogeochemical investigations of the Pidong Volcanic Crater Lake have also been carried out to constraint the chemical element sources into the Lake. Also, ^{40}Ar-^{39}Ar

IntechOpen

dating of the basalts of the Kassa volcanic field has revealed young ages spanning from 2.5, 1.97, 1.66, 1.39 to 1.34 Ma confirming the recent but short interval and multiple successive episodes of eruptions associated with volcanic activity in the region. Similar ages in the range of 2.1–0.9 Ma have been reported from basaltic rocks from the Benue Trough and a dolerite dyke from the region [1]. These same range of ages (of 2.83–0 Ma) have been obtained on basaltic rocks from the near-by Cameroon Volcanic Line (CVL) [1]. The Jos Plateau basalts vary in composition from basalt proper to trachyandesitic varieties of alkaline basalt magma series and some tholeiitic affinity [2]. Contrarily to the general notion that volcanoes are only associated with plate margins alone, the volcanoes in Nigeria are emplaced within the continent, associated with mantle Hot Spots. Records of gas emissions at Lakes Monoun and Nyos in Cameroon Republic in 1984 and 1986 respectively destroyed

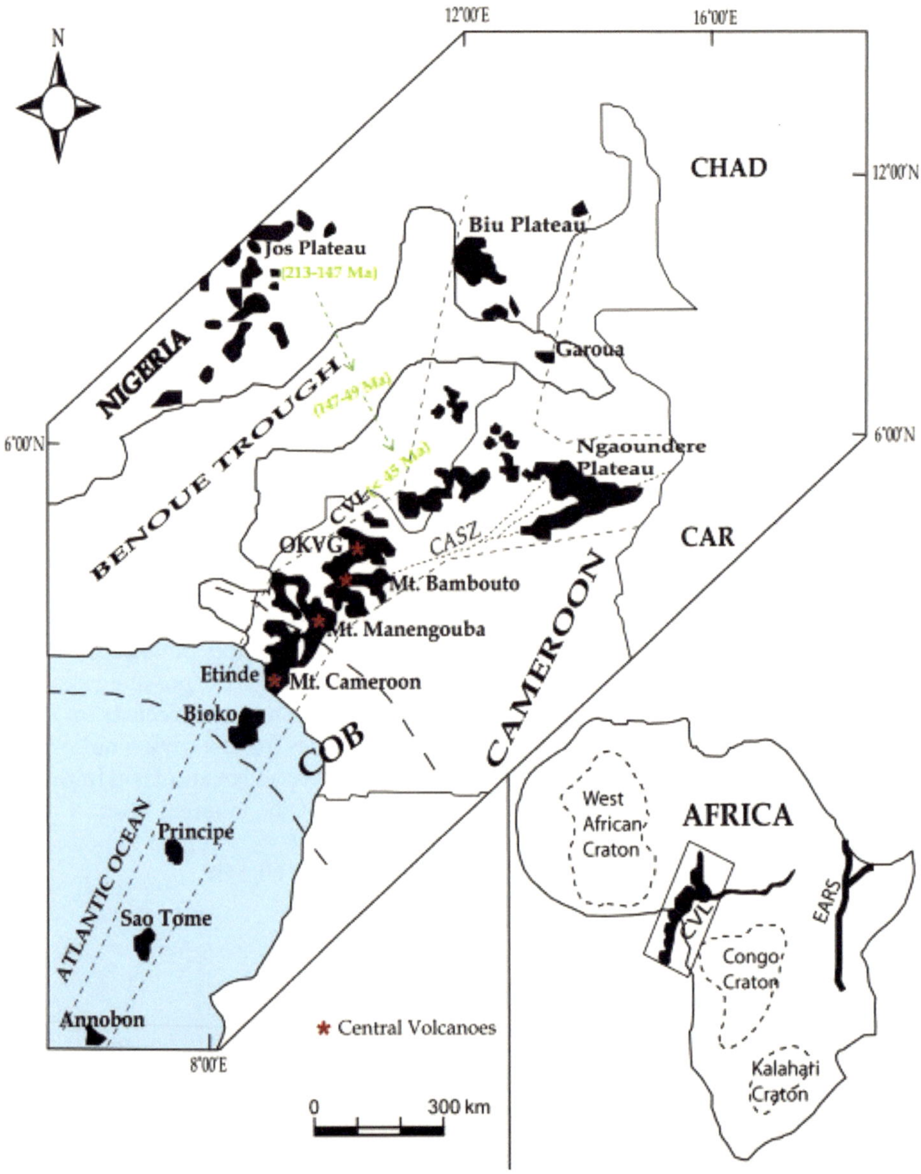

Figure 1.
Locations of the Jos and Biu Plateaux Volcanic Provinces, Nigeria.

lives and properties within 14 km radius to a large extent, affected some communities in Nigeria (Mambilla Plateau and Katsina-Ala River banks) [3]. This research attempts to provide answers as to whether it is possible that the dormant volcanoes in Nigeria with no recorded history of eruption could roar back to life? Some reported successive minor volcanic activities within the Cameroon Volcanic Line (CVL) (1909, 1922, 1954, 1959 and 1982) are pointers to the possibility of the reactivation of some dormant volcanoes in Nigeria.

2. Geologic settings

An understanding of the geology is important in studying natural hazards related to volcanism because the geology influences tectonic movement which leads to the disequilibrium of the ecosystem and other environmental components.

The Nigeria-Cameroon Volcanic Provinces lie within the Pan-African collision belt of West Africa (**Figure 1**). In Nigeria, the volcanic provinces (Jos Plateau, Biu Plateau, Benue Valley, etc.) are confined to the northeastern and central regions [4]. The volcanic provinces are characterized by numerous volcanic cones and lava flows consisting of alkaline olivine basalts together with less important trachyte and phonolite intrusive rocks [5].

Apparently, the large volcanic province of Nigeria suggests that volcanic activity during the Quaternary, was intense on the Jos Plateau, the Benue trough and along the northeast to southwest Cameroon line. Like the Jos Plateau volcanic regions, most of the main volcanic provinces such as the Bamenda Highlands, the Mambilla Plateau and the Adamawa Plateau are characterized by basement uplift [5].

The Cameroon Volcanic Line which extends from the Gulf of Guinea, Island of Annobon, Sao Tome and Principe and Fernando Po splits into two branches, one extending into northern Nigeria forming the Biu Plateau while the other extending eastward forming the Ngoude Plateau of Eastern Cameroun (**Figure 1**). Thus, in Nigeria, the volcanic provinces occur as relicts of volcanism in the form of scattered volcanic plugs and cones in some cases [5]. On the relatively uplifted Jos Plateau region, volcanic rocks are represented by volcanic cones and calderas [2].

Some of these cones generally rise only few hundred meters above the Plateau surface and are steep-sided with a central crater which may measure up to 450 m usually emerging at a breached in the crater wall. They are mainly built of basaltic scoria, volcanic ash, lava and with some variety of inclusions.

3. Materials and methods

3.1 Field mapping

Reconnaissance visits were carried out to past volcanic eruption sites where a careful and systematic mapping of each of the volcanoes visited were carried out so as to determine its nature, size and composition. Other information acquired was their physical surface weathering features, morphology, outcrop patterns and the extent of vegetation cover.

3.2 Petrographic and geochemical studies

Rock samples were collected for petrography, geochemistry and geochronology. Both petrographic and geochemical studies were done in the Department of Geology, University of Jos, Nigeria. ^{40}Ar-^{39}Ar dating on the basalts was done in

Netherlands, Geological Survey Laboratory. For quality control, a duplicate geochemical analysis on the same basaltic rocks was carried out at the University of Cardiff, Wales using the Inductively Coupled Plasma Optical Emission Spectrometry (ICP-OES).

3.3 Hydrogeochemical monitoring study

A constant monitoring of the Pidong Crater Lake through continuous water sampling 24–36 calendar months was done. The following physical parameters were recorded in the field using MT 806/pH/EC/TDS/Temp portable meter (pH, Temperature, Electrical Conductivity (EC), Total Dissolved Solids (TDS)). The water samples were collected in 100 ml polyethylene plastic bottles for cations and anion analysis. The sample for cations analysis was acidified with 0.1 M Nitric acid to prevent precipitating and bacterial growth. The following cations: Mg, Ca, Na, K, Cr, Ni, Co, Sc, V, Cu, Pb, Zn, Bi, Cd, Sn, W, Ma, As, Sb, Rb, Cs, Ba, Sr., Ga, Li, Ta, Nb, Hf, Zr, Y, Th, U, B, Fe & REEs were analyzed using ICP-MS method at Bureau Veritas Minerals Laboratory Limited, Canada while the anions: SO_4, Cl, HCO_3, NO_3, F, Br and PO_4 were carried out using colometry method at Maxxam Laboratory, Vancouver, Canada (subsidiary of Bureau Veritas Ltd).

3.4 Isotope study

Oxygen-18 ($\partial^{18}O$) and deuterium (∂^2H), and (∂^3H) and Carbon-14 (^{14}C) isotopes of the Pidong Crater Lake, Bwonpe Volcanic Spring and rainfall were analyzed at Activation Laboratory Ontario, Canada using cavity Ring down spectroscopy (CRDS) model L11 02-1 California, USA with V-SMOW standards with typical standards deviation for $^{18}O \pm 0.1\%$ and 1% for Tritium unit (TU).

4. Field investigations

Field investigations have revealed the existence of about 22 dormant volcanoes on the Jos Plateau region alone (**Table 1**) and are generally aligned in series of NNW–SSE trend [6–8] (**Figure 2**). None of these volcanoes have record of any activity in recent past [8, 9]. They are composed mainly of basaltic scoria and pyroclastic materials.

Tables 1 and **2** present the inventory of the dormant volcanoes on the Jos Plateau and the Biu Plateau, respectively. From the NNW end are the Ganawuri volcanic line, Hoss volcanic line, Panyam (Sura) volcanic line and Gu (Jiblik) volcanic line (**Figure 2**). The Ganawuri line comprises from the north to south of the Bum, Jal, and Kwakwi volcanoes. The Hoss volcanic line consists of two volcanoes; Miango in the north and Hoss volcano in the south. The Miango line consists of five volcanoes from the north to the south viz.: Rukuba, Miango north, Miango south, Vom and Kassa volcanoes. The Southern-most end members are namely the Panyam and Gu volcanic lines. The Panyam volcanic line consists of seven volcanoes aligned in a NNE–SSW trending directions along a hypothetical fracture line [6, 10]; Dai volcano (referred to as Wushik volcano), Amshel volcano (referred to as Kugol volcano), Dutsin volcano, Kerang volcano, Tingyaras volcano, Ampang volcano (referred to as Mufil volcano), Pidong Crater Lake.

The Kerang twin volcanoes are located at Kerang town and its environs in the present Mangu Local Government Area, about 120 km from Jos, Plateau State (**Figure 1**). The Kerang I (Dustin) volcano has a peak of 1456 m above sea level and a crater diameter of 300 m. The volcanic pile consists of ash, lapilli, scoria, pumices,

S. No.	Name/ locality	Coordinates	Estimated population of people at risk	Type of volcano	Diameter of crater	Elevation (ASL)	Materials deposited
1	Bum		200,000	Cone			Basaltic
2	Jal		150,000	Cone			Basaltic
3	Kwakwi		250,000	Cone			Basaltic
4	Miango volcano I	N09° 51′.365″; E008° 43′.961″	For 1 & 2 250,000	Cone	350 m	1297 m	Scoraceous basalt/ pyoclastics
5	Miango volcano II	N09 51. 000′; E008 44″. 191′		Cone	650 m	1303 m	‘’
6	Kassa volcanoes	Highest Peak: N09° 36′.119″; E008° 53′.521″	100,000	Cluster (6 overlapping volcanoes)	Average 300 m	Average 1342 m	Olivine basalt, scoria, tuff, breccia/volcanic bomb
7	Sha 1	N09° 10′. 543″;E008° 47′. 955″;	20,000	Dome	200 m	1310 m	Pyroclastics (granite fragments/lava)
8	Sha 2	N09°10′. 846″;E008° 48′. 05″;	10,000	Dome	200 m	1294 m	Weathered basaltic materials capped by iron concretions
9	Passakai	N09° 10′. 543″;E008° 47′. 955″	10,000	Dome	300 m	1375 m	Lateritized
10	Wushik (Lakas) volcano	N09° 24′ 165″; E009° 10′ 554″	10,500	Cone	250 m	1300 m	Scoria/ pyroclastics
11	Kogul (Nyeis) volcano	N09° 22′ 573″;E009° 11′ 068″	80,000	Cone	250 m	1250 m	Scoria/ pyroclastics
12	Kerang I	N09° 20′ 286″;E009° 11′ 643″;	I to IV put together 200,000	Cone	600 m	1400 m	Scoria/basaltic rocks with large phenocrysts of olivine, garnet and pyroxene
13	Kerang II	N09° 20′ 392″;E009° 11′ 502″;		Cinder Cone	1000 m	1450 m	Scoria/basaltic rocks with large phenocrysts of olivine, garnet and pyroxene
14	Kerang III volcano (Swan junction)	N09° 20′ 306″;E009° 10′ 561″		Cone	1000 m	1486 m	Scoria/ pyroclastics
15	Kerang IV	N09° 11′ 283″;E008° 12′ 547″		Cluster (with 2 craters)	1.5 km	1372 m	Pulverised basement and lava

S. No.	Name/ locality	Coordinates	Estimated population of people at risk	Type of volcano	Diameter of crater	Elevation (ASL)	Materials deposited
16	Pidong volcano	N09° 17′ 650″;E009° 12′ 312″	50,000	Crater Lake	700 m	1378 m	Scoria/ pyroclastics
17	Jiblik volcano	N09° 16′ 591″;E009° 16′ 890″	100,000	Cinder Cone	1 km	1228 m	Scoraceous basalt +garnet/ pyroclastics
18	Kagu volcano	N09° 13′ 901″; 008° 16′ 383″	50,000	Cone	1 km	1060 m	Scoraceous basalt/ pyroclastics
19	Katul volcano	N09° 11′ 264″;E009° 15′ 795″	5000	Cone	700 m	976 m	Scoraceous basalt/ pyroclastics
20	Lakdak		7000	cone			Scoraceous basalt/ pyroclastics

Table 1.
Inventory of the volcanoes of the Jos Plateau, Nigeria.

breccias, basalts, boulders, and pyroclastic materials of various sizes. The Kerang II (Kerang) volcano has three craters with a peak height of 1510 m above sea level with a crater diameter ranging from 600 m to 1 km (**Table 1**). This volcano (Kerang II) is the second largest volcano on the Jos Plateau compared to the Jiblik volcano which has a peak height of 1670 m above sea level. The volcanic pile of the Kerang II volcano is composed of ash, lapilli, scoria, breccias, bombs, basaltic boulders and pyroclastic materials (lapilli, granitic and lava flows).

The Pidong volcano (Maar) has a series of three craters and aligned along the general North-South trend of the Panyam Volcanic Line [10]. The Pidong Maar consists of a sequence of pyroclastic materials (mixture of large fragments of basement rocks/pyroclastics, scoria and ash) and indicative of violent eruption [9]. The Gu volcanic line consist of five volcanoes from NW to SW namely Jiblik, Kagu, Katul and Lagdak volcanoes. The volcanic cones are composed essentially of volcanic ash, lapilli, bombs, tuff agglomerates, basalts and scoria. Most of them occur as single cinder cones (like at Miango, Wushik, Kerang Swan junction, etc.) but rarely as clusters of two or more volcanoes (for example Jiblik, Kassa, Kerang twin volcanoes, Pidong, etc.)

Also, the relatively large sizes of some of these volcanoes (Miango, Kassa, Jiblik, Kerang, etc.) suggest that quite a large volume of ejecta materials were spewed out covering quite a large landmass (valleys and low-lying plains) as lava flows. Lava flows apparently from the Jiblik volcanic can be traces to several kilometers south of the Jos Plateau escarpment. If any of these volcanoes erupt today with the same intensity and volume presumed, a large expanse of land would be buried and about 2 million people living around these volcanoes are potentially at risk.

The volcanoes of the Biu Plateau (**Table 2**) present similar physical and petrographic characteristics as those of the Jos Plateau region. The volcanoes also form near linear alignments from the north to the south and extend right through the low-lying Basement complex into the sedimentary formations of the Benue valley (Garkida-Gombi-Song areas in Adamawa State). The volcanoes are simple and never in clusters, but with very large craters of greater than 1 km, (Caldera). The volcanoes extruded directly the basement rocks and therefore are of lower

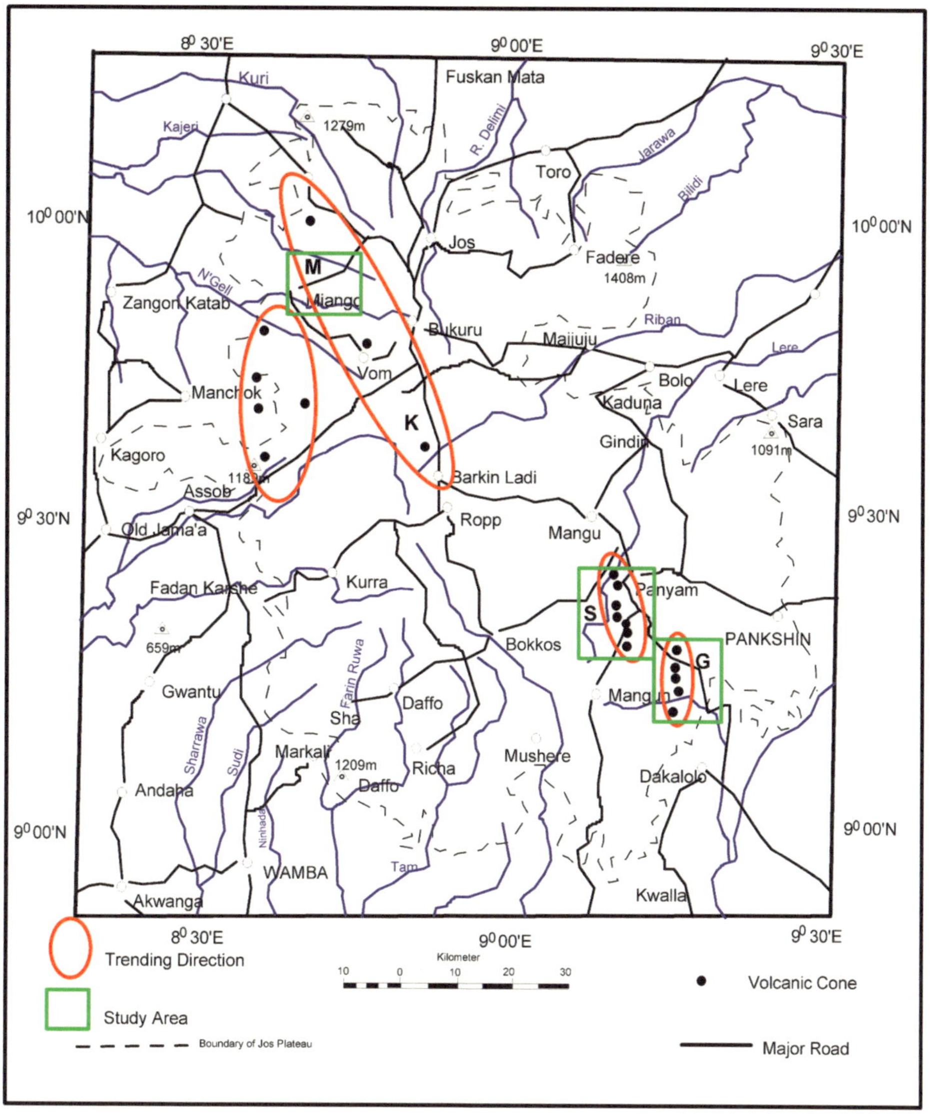

Figure 2.
Geological map of the 22 dormant volcanoes of the Jos Plateau showing sample locations. M = Miango volcanoes, K = Kassa volcanoes, S = Sura volcanoes, and G = Gu volcanoes.

altitude above sea level relative to those of the Jos Plateau, which extruded the already high-level Younger Granite bodies.

Also, unlike the volcanoes of the Jos Plateau, those of Biu Plateau are constituted by a large volume of volcanic ash and pyroclastic materials (for example, at Gwamya, Tilla Crater Lake, Gadam, Batadeka, Buratai, Katla volcanoes, etc.) (**Table 2**), suggesting that there was a tremendous spewing of ejecta materials (ashes and gases) into the atmosphere prior to the violent eruptions and/or in-between the eruptions.

5. Geochemical results and interpretations

The major and trace elements geochemical compositions of the basaltic rock samples collected from the various volcanic cones in the Jos Plateau volcanic

S/ No.	Name/ locality	Coordinates	Estimated population of people at risk	Type of volcano	Diameter of crater	Elevation (asl)	Materials deposited
1	Tasha Village	N10°17′388″ E011° 24′406″	5000	Cluster of 3 volcanic cones.	50 m	426 m asl	Massive basaltic rocks
2	After Tasha Village	N10° 16′.661″ E011° 27′.143″	5000	Dome		427 m	Massive/ vesicular basaltic rocks
3	Tagwaye (twin) volcanoes (In Kwaya Kusar LGA of Borno State)	N10° 30′.530″ E011° 30′.402″	15,000	2 Cones	200 m	512 m	Olivine basalts/ Agglomerates.
4	Gadam Volcano (Kwayar Kusar LGA, Borno State)	N10° 31′.951″ E011° 53′.485″	30,000	Cone	250 m	479 m. asl	Olivine Basalt
5	Location 5	N10° 32′.876″ E011° 57′.751″	Inhabited	Plug		683 m	Basaltic boulders/ Agglomerates/ tuff
6	Location 6	N10° 30′. 279″ E011° 50′ 726″	Inhabited	Dome		826 m	Vesicular Basaltic boulders/ Agglomerates/ tuff
7	TUM	N10° 36′. 254′ E012° 06′ 564″	10,000	Dome		643 m asl	Scoraceous basalt with olivine/ zeolite/ Columnar basalt.
8	Wakama (a) (BCG Village)	N10° 36′.472′ E012° 07′ 126″	2500	Cone	1 km	677 m	Scoraceous basalt
9	Wakama (b)	N10° 37′.211′ E012° 08′ 886″	2500	Caldera	10 km	701 m	Scoraceous basalt
10	Gwamya Volcano	N10° 33′.496″ E012° 06′. 341″	6000	Cone	500 m	752 m	Scoraceous basalt/ pyroclastic pile
11	Tilla Volcanic Hill	N10° 32′.549″ E012° 08′.477″	2500	Cone	500 m	910 m	Scoraceous basalt/ pyroclastic pile
12	Tilla Crater Lake	N10° 32′.336″	10,000	Caldera	2 km	751 m	Scoraceous basalt/

S/ No.	Name/ locality	Coordinates	Estimated population of people at risk	Type of volcano	Diameter of crater	Elevation (asl)	Materials deposited
		E012° 07′. 945″					pyroclastic pile
13	Versu Volcano	N10° 39′.759″ E012° 08′.457″	3000	Seasonal Crater Lake	700 m	782 m asl	Olivine basalt
14	Dragna Volcano	N10°27′.132″ E012° 05′.820″	5000	Caldera filled up by collapsed materials.	3 km	750 m asl	Weathered scoraceous basa Basalt
15	Marama Volcano	N10° 27′.746″ E012° 09′.093″	150,000	Cone	1 km	735 m	Scoraceous basalt/ pyroclastics
16	Gwaram volcanic hill	N10° 39′.214″ E012° 08′.092″	8000	Cone	1 km	942 m	Boulders of scoriaceous basalt
17	Batadeka (i) Volcanic hill	N10° 41′. 673″ E012° 07′.707″	3000	Cone	700 m	884 m	Weathered scoriaceous basalt
18	Batadeka (ii) Volcanic hill	N10° 42′. 067″ E012° 07′.542″	3000	Cone	1 km	1053 m asl	Weathered scoriaceous basalt
19	Kwatla Crater Lake	N10° 42′.117″ E012° 06′.433″	Inhabited	Seasonal Crater Lake	1.5 km	752 m asl	Weathered scoriaceous basalt
20	Maldau	N10° 43′.500 E012° 06′.979″	Inhabited	Cone	350 m	907 m asl	
21	Buratai Volcanic Hill	N10° 53′.926″ E012° 02′.800′	5000	Dome	700 m	718 m asl	Weathered scoriaceous basalt/ pyroclastics
22	Kona Uku Volcanic hills	N10° 49′.814″ E012° 07′.062″	10,000	3 Cones clustered together	2 km	879 m asl	Weathered scoriaceous basalt/ pyroclastics
23	Dutsen Kura (Bogur) Volcanic hill	N10° 49′.313″ E012° 05′.978″	5000	Seasonal Crater Lake	1.2 km	1011 m asl	Weathered scoriaceous basalt/ pyroclastics
24	Kukuwa (Gabai lga Yobe State)	N11° 06′.387″ E011° 53′.689″	5000	Plug	200 m	429 m asl	Columnar basalt

S/ No.	Name/ locality	Coordinates	Estimated population of people at risk	Type of volcano	Diameter of crater	Elevation (asl)	Materials deposited
25	Kukuwa II	N11° 06′. 443″ E011° 53′. 534″	5000	Several Plugs	200 m	435 m asl	Columnar basalt
26	Kurara Volcanic Hill (Garkida junction Adamawa State)	N10° 22′.460″ E012° 34′.284″	50,000	Cone	1 km	661 m asl	Weathered scoriaceous basalt/ pyroclastics
27	Song (Song-Gombi Road)	N9° 51′.036″ E012° 36′.383″	50,000	Cone	700 m	477 m	Weathered scoriaceous basalt/ pyroclastics
28	Song (Hawul Mountains)	N9° 49′.504″ E012° 37′.155″	10,000	4 different cones aligned N-S	250 m each	420-1000 m asl	Weathered scoriaceous basalt/ pyroclastics

Table 2.
Inventory of the volcanoes of the Biu Plateau, Nigeria.

province are presented in **Tables 3** and **4**. The volcanic cones situated at the northern end of the volcanic line here referred to as the north-western group are represented by Miango (M1 & M2) and Kassa (K1–5), while the south-eastern end group are represented by Jiblik (G1 &G2), Tingyaras (S1), Ampang (S2), Pidong (S3), Wulshik (S4), Kugol (S5) and Kerang (S6).

The rocks display similar SiO_2 wt% contents (44.84–50.06 wt%) for the north-western group of volcanoes (M1&2 and K1–5) and 45.26–46.25 wt% for S1–6 and 49.69 wt% for the Jiblik volcano (G1). However, the sample G2 from Jiblik with high SiO2 content of 64.21 wt% is exceptionally acidic and does not seem to be a basalt. Many of the rocks from the Kassa volcanoes (K1, K3, K4, and K5) display the highest SiO_2 content (46.99–50.06 wt%) as opposed to those from Miango volcanoes (M1 &M2) and the southern volcanoes (S1–6) whose SiO_2 contents are typical of a normal basalt (45.26–46.25 wt%).The Miango and the Kassa volcanoes display higher Fe_2O_3 contents (12.33–12.61 and 10.42–11.35 wt% respectively). The southern group (G1 & S1–6) displays lower Fe_2O_3 content; 9.83 wt% for the Jiblik volcano (G1); and a relatively higher but similar concentrations of between 11.25 and 11.92 wt% for the Panyam Volcanic line members (S1–6).

In general, the Al_2O_3 contents of all the basalts from the different volcanic cones are significantly high but vary narrowly inter/intra the volcanic cones (13.87–18.07 wt % for the north-western group and 12.41–18.07 wt% for the southern group (G1 & S1–6)). In terms of the MgO content, the southern group (S1–6) presents relatively higher values of between 10.71 and 12.58 wt% as against an average of 8.00 wt% for the northern group (M1–2 & K1–5). The CaO, Na_2O, K_2O and TiO_2 contents for all the volcanoes vary narrowly; averagely 8; 3; 1.5 and 2 wt%, respectively.

Their Al_2O_3/TiO_2 ratios vary narrowly and could be on that basis be subdivided into two groups; those with ratios between 5.53 and 5.69 and then those between 6.39 and 6.62 for the entire volcanic line (N-S) (**Figure 3**). This is further a

Analyte symbol unit	SiO_2 (%)	Al_2O_3 (%)	Fe_2O_3 (%)	MnO (%)	MgO (%)	CaO (%)	Na_2O (%)	K_2O (%)	TiO_2 (%)	P_2O_5 (%)	LOI (%)	Total (%)	Al_2O_3/TiO_2	CaO/TiO_2	Al_2O_3/CaO
M1	44.84	15.02	12.33	0.166	8.50	8.86	4.09	1.88	2.668	0.76	<0.01	98.83	5.63	3.32	1.70
M2	43.80	15.01	12.61	0.166	8.54	8.32	3.40	1.87	2.727	0.74	1.5	98.68	5.50	3.05	1.80
G1	49.69	14.90	9.83	0.142	8.10	7.07	3.34	1.70	1.659	0.47	1.33	98.81	9.03	4.26	2.11
G2	64.21	14.74	5.60	0.087	2.37	3.91	3.29	3.54	1.124	0.43	0.96	100.3	13.11	3.48	3.77
S1	44.97	12.41	11.51	0.166	12.85	9.43	3.13	1.62	2.213	0.59	0.03	98.92	5.61	4.26	1.32
S2	44.77	12.48	11.34	0.165	12.74	9.39	2.96	1.50	2.194	0.57	0.47	98.58	5.69	4.28	1.33
S3	45.75	16.00	11.92	0.161	6.34	8.16	3.58	2.21	2.849	0.70	0.99	98.66	6.62	2.86	1.96
K1	46.99	13.99	11.36	0.630	8.36	9.47	3.74	1.53	2.176	0.60	0.52	98.95	6.43	4.35	1.48
K2	50.06	14.23	10.83	0.149	7.16	8.9	3.50	1.22	2.158	0.37	<0.01	98.57	6.59	4.12	1.60
K3	48.02	18.07	10.42	0.164	3.30	7.74	5.28	2.80	2.516	0.84	0.12	99.27	7.18	3.08	2.33
K4	47.49	14.17	11.35	0.167	7.81	9.29	3.82	1.58	2.211	0.64	0.20	98.74	6.41	4.20	1.53
K5	47.05	13.87	11.28	0.160	8.17	9.36	3.55	1.59	2.17	0.64	0.67	98.52	6.39	4.31	1.48
S4	45.26	13.52	11.34	0.162	10.71	9.47	3.28	1.70	2.397	0.62	0.32	98.78	5.64	3.95	1.43
S5	44.77	12.82	11.40	0.166	11.50	9.58	2.83	1.65	2.318	0.61	1.08	98.73	5.53	4.13	1.34
S6	46.27	13.02	11.25	0.163	11.24	9.21	3.51	1.69	2.262	0.69	0.23	99.35	5.76	4.07	1.41

Table 3.
Major element compositions (in weight percentage) of volcanic rocks from the Jos Plateau Volcanic Province.

Analyte symbol unit	Ba ppm	SR ppm	Y ppm	Sc ppm	Zr ppm	Be ppm	V ppm	Ba/Sr	Zr/Y
M1	619	840	25	19	234	2	161	0.74	9.36
M2	865	799	26	19	239	2	153	1.08	9.19
G1	518	637	18	15	205	2	122	0.81	11.39
G2	1215	723	26	12	328	3	121	1.68	12.62
S1	530	654	22	23	193	2	185	0.81	8.77
S2	554	646	21	22	190	2	178	0.86	9.05
S3	794	1098	24	15	228	2	174	0.72	12.00
K1	571	782	25	18	177	2	161	0.73	7.08
K2	551	460	43	20	155	2	171	1.20	3.06
K3	820	983	26	7	258	3	155	0.83	9.92
K4	672	758	38	19	176	2	164	0.89	4.63
K5	523	1095	24	18	177	2	157	0.48	7.36
S4	569	748	23	21	212	2	178	0.76	9.22
S5	585	697	23	23	204	2	180	0.84	8.84
S6	688	766	23	20	215	2	172	0.90	9.35

Southeastern volcanoes: G1—basalt of Jiblik volcano; S1—basalt of Timjagha'as volcano; S2—basalt of Ampang volcano; S3—basalt of Pidong volcano; S4—basalt of Wushik volcano; S5—basalt of Kogul volcano; S6—basalt of Kerang volcano. Northwestern volcanoes: M1—basalt of Miango North volcano; M2—basalt of Miango South volcano; K1—basalt of Kassa volcano; K2—basalt of Kassa volcano; K3—basalt of Kassa volcano; K4—basalt of Kassa volcano; K5—basalt of Kassa volcano.

Table 4.
Trace elements compositions (in ppm) of the basaltic rocks from the Jos Plateau Volcanic Province.

reflection of the variations in Al_2O_3 contents of the basalts since TiO_2 contents remain relatively constant. Similarly, their CaO/TiO_2 and Al_2O_3/CaO ratios vary from 2.6 to 4.31 and 1.34–1.60 respectively. Such narrow difference in these ratios is expected from a low degree of magmatic differentiation of the same parent material by partial melting process (**Figures 3** and **4**).

Similarly, the rocks present subtle variations in incompatible element ratios Ba/Sr. and Zr/Y (0.73–0.90 and 8.77–9.35, respectively); all supportive of their subjection to low degree of magmatic differentiation and/or similar source.

5.1 Silica versus major oxides correlation plots

In the SiO_2 versus MgO wt% plot, the southern volcanoes overwhelmingly display higher MgO contents as opposed to the lower values for the northern group of volcanoes (**Figure 5c**). It is expected that the Kassa volcanoes (K1–5) which are more differentiated (by their higher SiO_2 contents) than the others to present lower MgO contents but instead display similar MgO contents. This scenario is true of their Fe_2O_3 contents. However, there is a weak negative correlation between Fe_2O_3, MgO, TiO_2 and MnO versus silica indicating a progressive decrease of these oxides with differentiation (Miango-Southern group-Kassa) corresponding to compositional variations related to the removal of different proportions of olivine/pyroxenes from the melt as it becomes more felsic (**Figure 6a–d**).

The Kassa volcanoes in the northern group and those of the southern group exhibit higher CaO contents compared to lower CaO values at Miango (M1&M2) and also of the northern group. The observed high CaO content suggest the

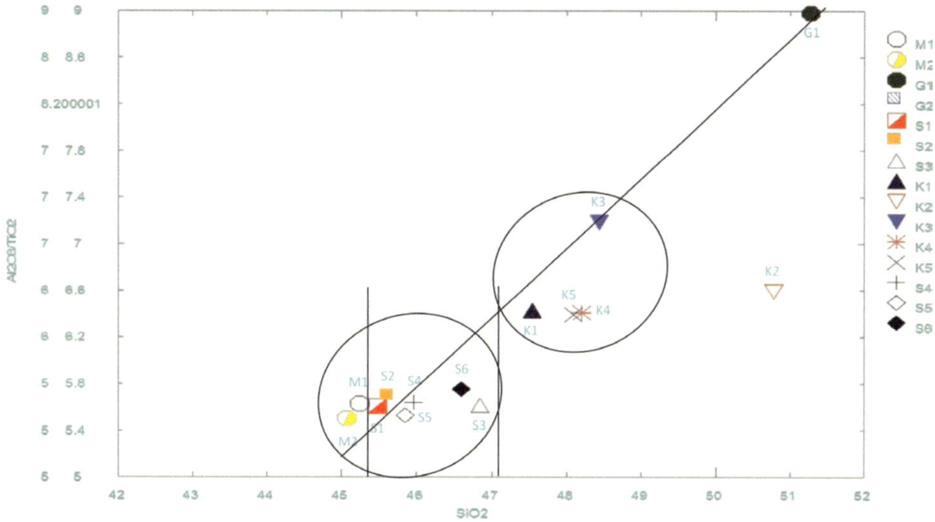

Figure 3.
Plot of SiO_2 *versus* Al_2O_3/TiO_2 *ratios.*

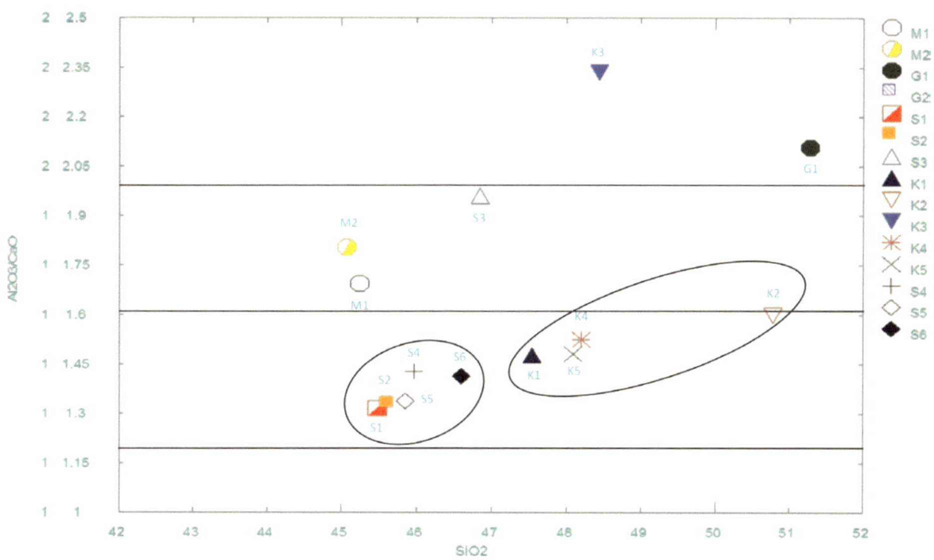

Figure 4.
Plot of SiO_2 *versus* Al_2O_3/CaO *ratios.*

recrystallization of clinopyroxenes in the early stages of crystallization. In general, there is a positive correlation depicting a progressive increase of CaO with differentiation from Miango to the southern group to Kassa (**Figure 5a–f**). This increase is not visible at the level of one volcano but several of them put together. The high Fe_2O_3, MgO and CaO could be as a result of the bulk crystallization of olivine, pyroxene and plagioclase during the early stages of differentiation. The progressive increase in the contents of Al_2O_3 corresponds with the increase in the alkaline metals content (Na_2O + K_2O) suggesting the crystallization of plagioclase with increased degree of differentiation. The alkaline oxides (Na_2O + K_2O) are correspondingly highest in the most differentiated rocks (K1–5) (**Figure 6d**). In a silica versus total alkali diagram (**Figure 7**), the rocks fall within the alkaline field and are therefore classified as predominantly alkaline basalts. Only a few rocks fall in the

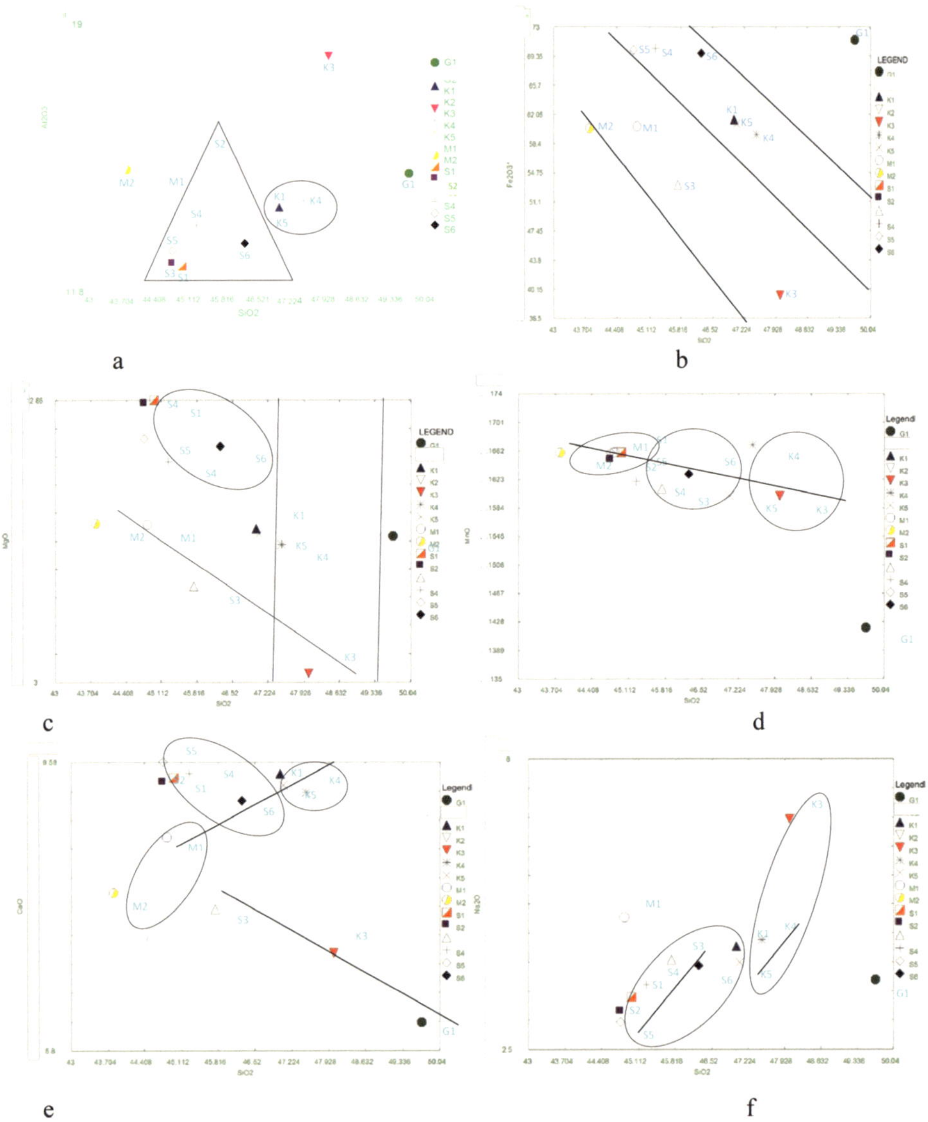

Figure 5.
(a–f) Plots of SiO_2 versus major oxides of the volcanic rocks of the Jos Plateau Volcanic Province.province.

sub-alkaline field which could be of tholeiitic character (high silica and Fe Contents). Furthermore, in a log (Zr/TiO_2) versus SiO_2 diagram (**Figure 8**), the bulk of the rocks fall in the alkaline field reaffirming their alkaline nature (**Figure 9**).

In respect to their Mg#, it varies very narrowly within the individual volcano signifying a subtle degree of differentiation (partial melting) and thus reflecting a low degree of magmatic differentiation and the consequent subtle compositional variations observed. However, for the entire volcanoes put together, the Mg# vary significantly from 27 to 56 suggesting formation of the rocks by fractional crystallization at larger scale. The southern volcanoes (S1–6) have the highest Mg# relative to those of the northern group (K1–5 and M1 & M2) (**Figure 10a–d**). When the rocks are plotted in Mg# versus Fe_2O_3 diagram (**Figure 10b**), a positive correlation is obtained indicating a progressive decrease in Fe_2O_3 with increasing degree of magmatic differentiation. The relatively similar Ba/Sr. and Zr/Y ratios for these rocks but with progressive decrease in Mg# lends credence to their derivation from

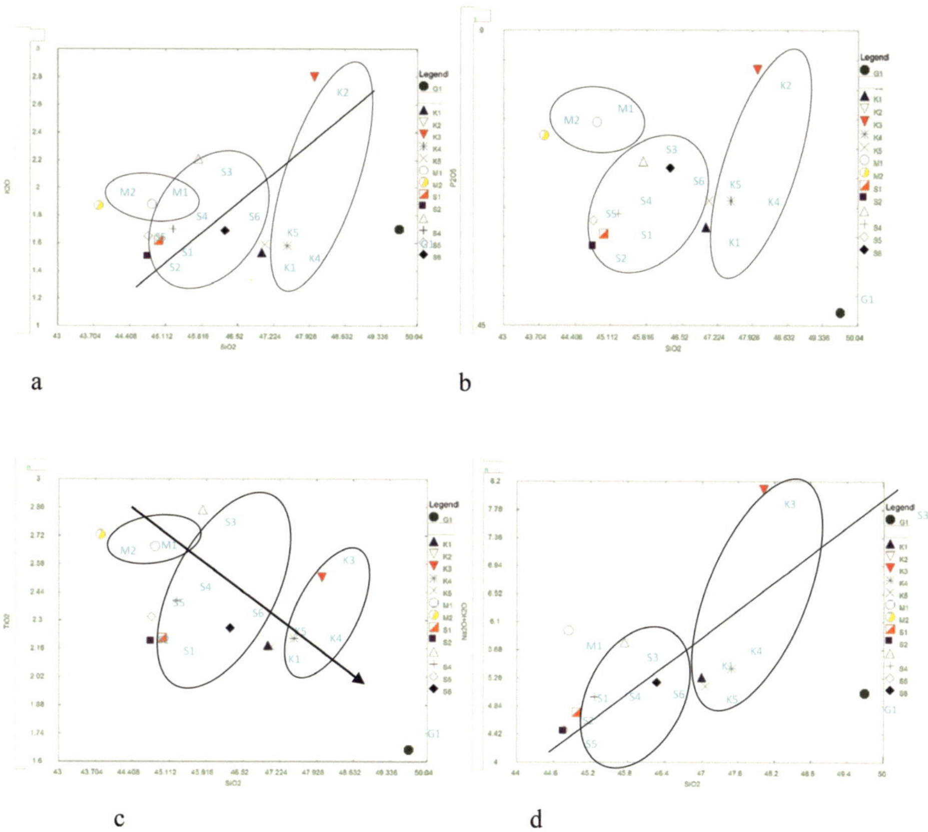

Figure 6.
(a–d) Plots of SiO_2 versus major oxides of the volcanic rocks of the Jos Plateau Volcanic Province.

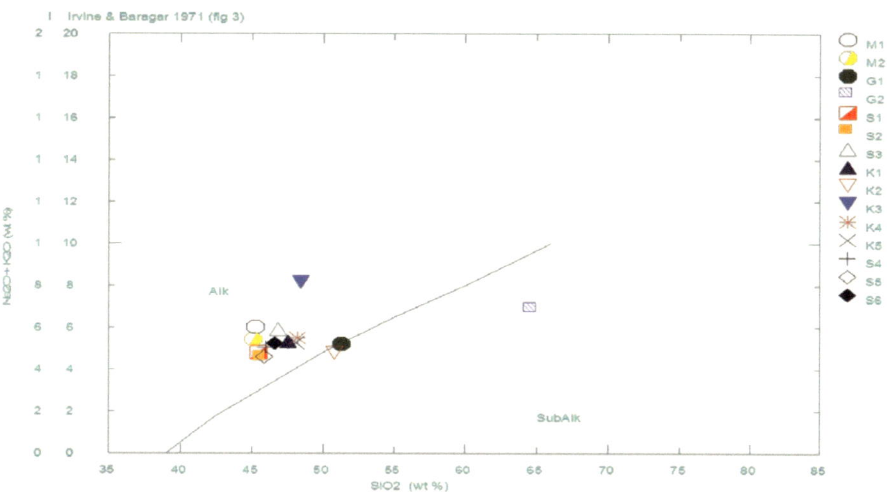

Figure 7.
Silica versus Total Alkali diagram [11] of the volcanic rocks of the Jos Plateau (Alk = alkaline and Subalk = subalkaline).

the same magma reservoir by differentiation. It appears that the rocks with the highest Mg# (samples S1–6) present compositions that are close to that of the parent materials since the magma did not suffer high degree of differentiation giving rise to a variety of rocks. The subtle variation of Mg# and the incompatible

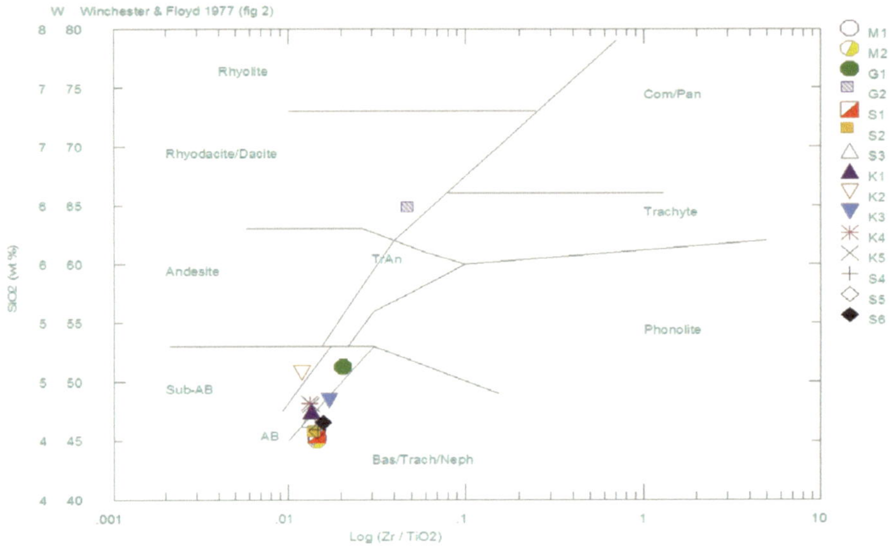

Figure 8.
Plot of Log (Zr/TiO_2) ratio versus SiO_2 [12] of the volcanic rocks of Jos Plateau Volcanic province (Sub-AB = subalkaline basalts, AB = alkaline basalts).

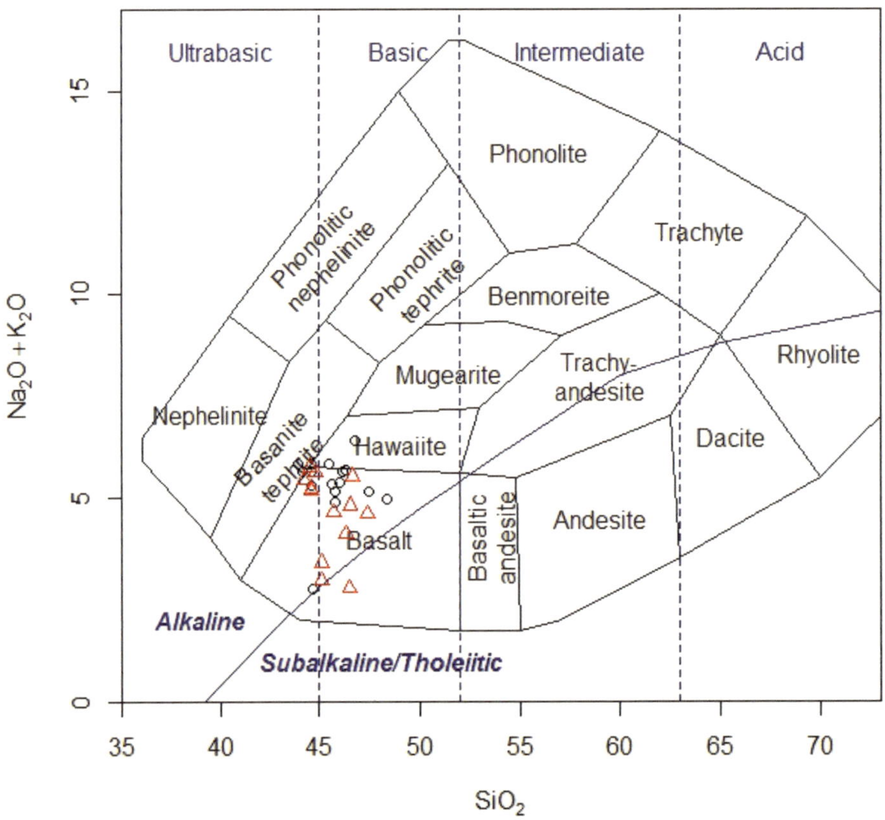

Figure 9.
SiO_2 versus Na_2O + K_2O classification diagram of basalts of the Jos Plateau volcanoes [13].

DOI: http://dx.doi.org/10.5772/intechopen.84253

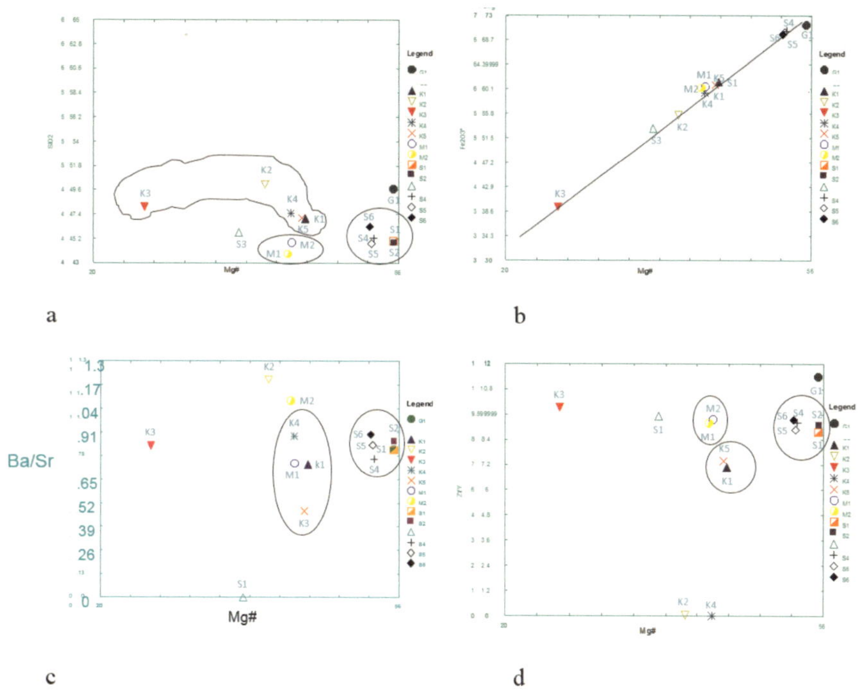

Figure 10.
(a–d) Plot of some selected major oxides versus Mg# for the volcanic rocks of the Jos Plateau volcanic province.

element ratios highlighted above are supportive of the derivation of these basalts by partial melting process of a magma from the same source.

5.2 Tectonic environment of emplacement

The geochemical data plotted in the ternary diagram of Ti/100-Zr-Y*3 [12] show clearly that they were majorly emplaced within the continental crust (**Figures 11** and **12**). This fact distinct these basalts from those of the Island arc and the Mid Oceanic Ridge.

5.3 Incompatible elements spidergraph

The incompatible elements when plotted in a spidergraph normalized to Chondrites in comparison with OIB (**Figure 13**) display a relatively similar pattern with slight enrichment in their incompatible elements. These characteristic features are typical of most alkali basaltic suites derived from a deeper mantle source akin to that of the OIB [12, 14, 16–19].

5.4 Ar^{40}-Ar^{39} dating

A sample by sample result is presented in **Table 5**. The Ar/Ar ages span between 1.3 and 2.5 Ma, confirm the earlier K-Ar ages of 2.1 and 1.9 ± 0.31 Ma reported by [1] on dolerites on the Jos Plateau. The short interval in the radiometric ages suggests volcanic eruptions occurred at discrete times, separated by short periods of non-activity at a mean age average of 0.55 Ma (CN3 = 2.500 ± 0.318 Ma and CN5 = 1.970 ± 0.173 Ma). This long period must have been dominated by profound erosion. The considerable long-time difference from the oldest to the youngest eruption suggests that there was relatively steady magma source overtime.

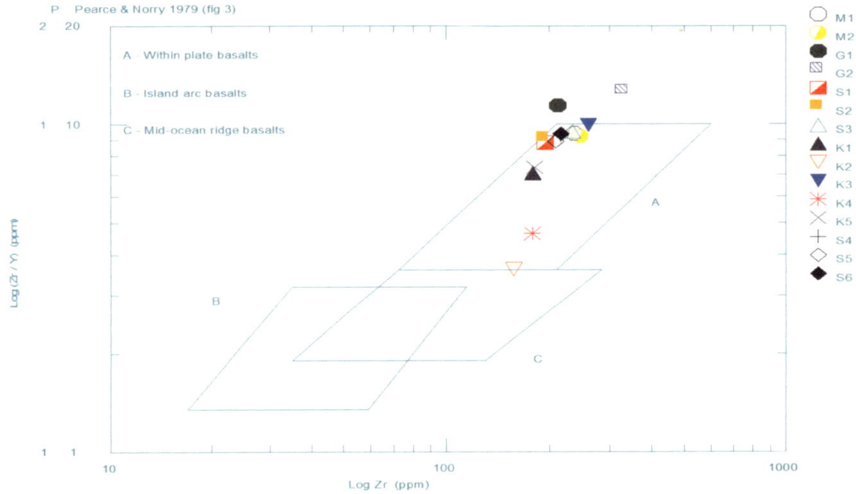

Figure 11.
Log Zr (ppm) versus Log (Zr/Y) (ppm) diagram [14] volcanic rocks of the Jos Plateau Volcanic province.

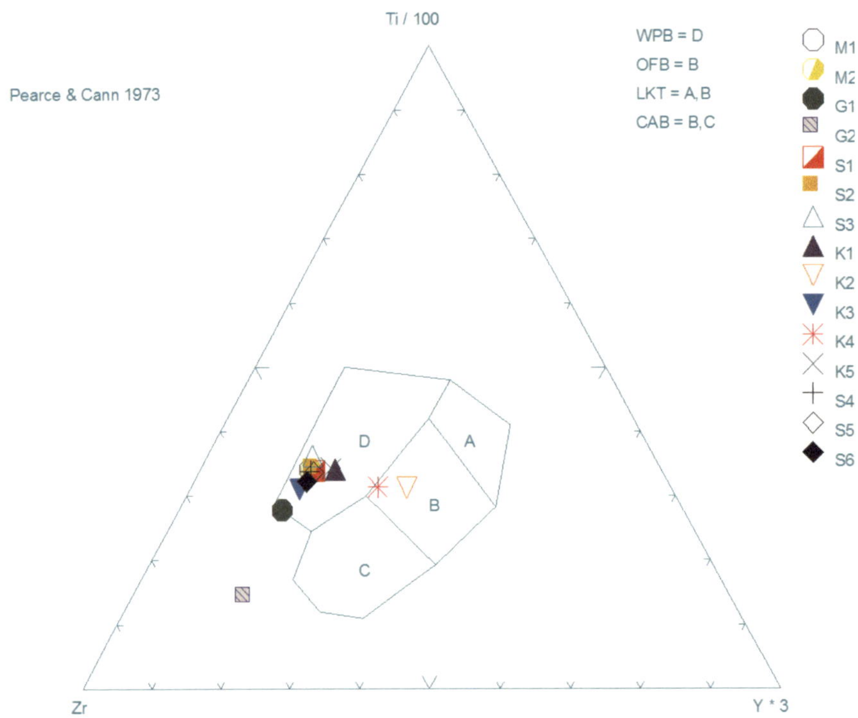

Figure 12.
*Ternary diagram of Ti/100 – Zr – Y*3 [12] for the volcanic rocks of the Jos Plateau Volcanic province. WPB = D, within plate basalts; OFB = B, ocean flood basalts; LKT = A,B low potassium tholeiitic basalts; and CAB = D,C, calc-alkaline basalts.*

5.5 Hydrogeochemistry of the Pidong Crater Lake

Comparative hydrogeochemical parameters of previous study [7] and this present study is presented in **Table 5**. It shows clearly that pH and alkalinity have decreased overtime from 9.35 to ≥7.0 and from 335 to 145 mg/l, respectively as well

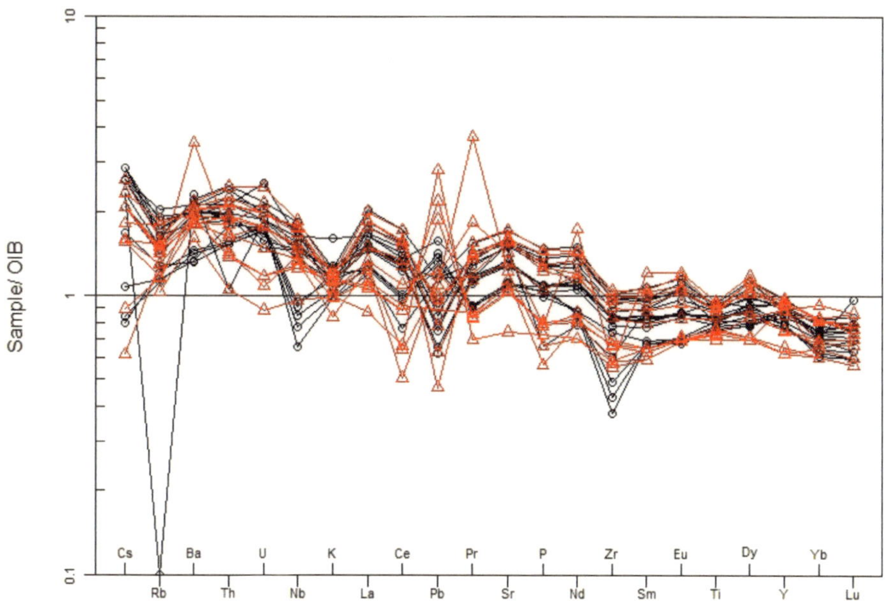

Figure 13.
Spidergraph plot of incompatible element compositions of basaltic rocks from the Plateau volcanoes normalized to OIB [15].

as increase in the concentrations of Cl from 2.5 to 5.6 mg/l; SO_4 from <0.33 to 1.05 mg/l (**Table 5**).

5.5.1 Major element concentration/distribution

In general, the major element concentrations in the Pidong Lake decrease in the order Mg > Ca > K > Na (**Table 5**). The highest concentrations of Mg (30 to ≤40 mg/l), Ca (21–25 mg/l), K (15 mg/l) Na (8–11 mg/l) are observed during the dry season (January–April) while lowest concentrations of Mg (16–25 mg/l), Ca (10–15 mg/l), K (4–7 mg/l) in the rainy season (August–October). The intermittent change of color of the Lake from clear blue to brown has been attributed to the increase in Fe concentrations into the Lake.

5.5.2 Rare earth element (REEs) concentrations

Figure 14 presents the REE patterns from the Crater Lake normalized to Chondrites. The REE concentrations are impoverished relative to Chondrite values (<1 × Chondrite). There are significant variations in the LREE (0.03–0.18 × Chondrite for La) relative to HREE (Gd-Lu). An important characteristic of the spectra is the similarities between the LREE patterns indicating similar source. The slight enrichment in LREE must have been influenced by fluid percolation through the host crustal materials (host granite basement) rich in these elements.

5.5.3 Anion concentrations (SO_4, Cl, HCO_3, NO_3, F, Br, and PO_4)

The major anion concentrations from the lake vary in concentrations in the order of -HCO_3 > Cl > SO_4. The highest concentration of in the LREE range from <0.5 to 5.5 mg/l and 0.0675 to 0.0321 mg/l, respectively, and were observed during

Period	pH	Temperature (°C)	EC (µs/m)	Mg (mg/l)	Ca (mg/l)	Na (mg/l)	K (mg/l)	Zn (mg/l)	Fe (mg/l)	Cu (mg/l)	Al (mg/l)	SO_4 (mg/l)	Cl (mg/l)	Alkalinity ($CaCO_3$) (mg/l)	F (mg/l)
Patterson (1986) report	9.35	28.5	410	36.4	13.65	12.8	25.8	0.065	0.27	<0.004	—	<0.33	2.5	335	—
Present study (2013–2015)	7.17	26.1	299	29.68	39.67	6.89	11.39	0.064	0.612	0.0039	0.456	1.03	5.67	141.5	0.249

Table 5.
Comparative hydrogeochemical data of 1986 and the present study.

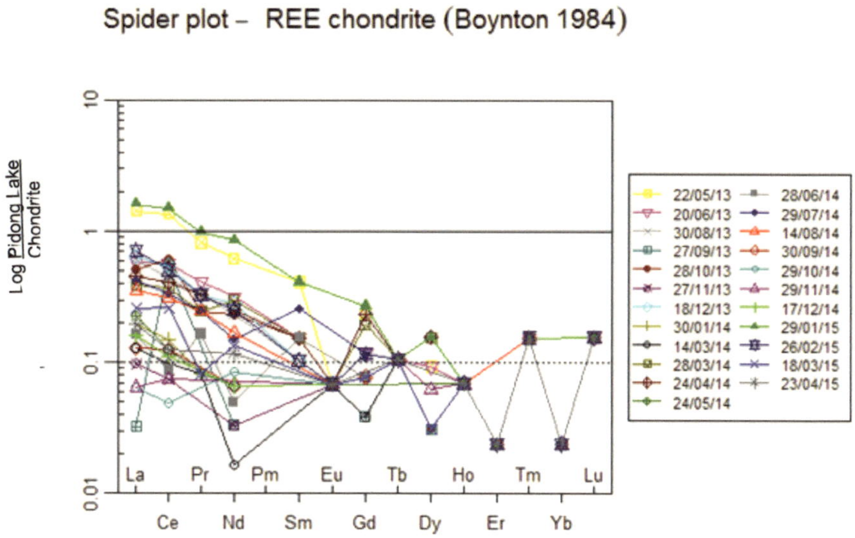

Figure 14.
Spidergraph of rare earth elements (REEs) of Pidong Lake normalized to Chondrite [20].

the dry season (October–April). The concentrations of HCO_3, Cl and F could have been influenced by interacting percolating water and host rock chemistry.

The increase in the concentrations of SO_4^{2-}, Cl^- (<0.33–1.03 and 2.5–5. 67 mg/l, respectively) as well as the decrease in Alkalinity and pH from 335 to 141.5 mg/l and 9.35 to ≈7.00, respectively. These variations in concentrations suggest a possible input of magmatic/fumaroles percolating upwards to shallow groundwater aquifers coming in contact with meteoric water.

The oxygen and hydrogen isotope composition and plot of $\partial\,^{18}O$ and $\partial\,^{2}H$ relative to SMOW line shows that the Pidong Crater Lake is of meteoric origin.

6. Conclusion

i. This work presents and affirms the existence of relicts of past volcanic eruptions (dormant volcanoes) in Nigeria (Jos and Biu Plateaux).

ii. That these volcanic edifices were previous eruption sites suggest they are potential eruption sites. The proximity of these volcanic edifices to those of the Cameroon volcanic line, which has witnessed a series of volcanic activity, is worrisome.

iii. The mafic lavas in both the Jos and the Biu Plateaux volcanic provinces display geochemical compositions that are characteristic of alkaline basalts

iv. Petrologically, the basaltic rocks display compositions varying from basalt proper to basanite-tephrite-hawaiite, emplaced within the host crustal rocks (Basement rocks).

v. The chain of volcanoes evolved from the same basalt parent magma by crystal fractionation each cluster derived by partial melting of the same residual parent magma.

vi. Recent $^{40}Ar/^{39}Ar$ dating of basalts from overlapping volcanic cones from the Jos Plateau province (ranging from 2.5 to 1.34 Ma) confirms the Quaternary age (Pleistocene epoch) of emplacement for these volcanoes and the intermittency of eruptions inter and intra the volcanic line in the provinces.

vii. That these volcanic edifices were previous eruption sites suggest they are potential eruption sites. The proximity of these volcanic edifices to those of the Cameroon volcanic line, which has witnessed a series of volcanic activity, is worrisome. The volcanic eruption of Mount Vesuvius in Italy in the year 2004 known to have been dormant since 24 AD is a clarion call. Furthermore, the recent volcanic eruption in Iceland April 15, 2010 after 200 years of silence should increase the worry. All these information calls for further comprehensive work on these dormant volcanoes for risk assessment.

viii. The minor intermittent fumarolic activities observed within the Pidong Crater Lake marked by the change in the water color from the normal bluish to brown-red color, (due to Fe-input) call for more comprehensive investigation of the volcano and further strengthen the idea of the possibility of the volcano roaring back to life.

Acknowledgements

This study is an on-going project on volcanoes in Nigeria, being embarked upon by three PhD scholars. It is being supported by STEP-B World Bank funding. Also, the Centre for Geodesy and Geodynamics, National Space Research and Development Agency (NASRDA) is well appreciated for providing funds for the field work.

Conflict of interest

We declare that there are no "conflict of interest."

Notes/thanks/other declarations

The authors are deeply indebted to the Geological Survey of Norway for dating the rocks. We express our deep thanks to anonymous reviewers of this manuscript, who have provided constructive criticisms which has improved the quality of the initial manuscript. Our profound gratitude goes to our project students who assisted in the sample collection.

Appendices and nomenclatures

Alk	alkali basalt
ASL	above sea level
Ma	million (10^6) years
MORB	mid-ocean ridge basalt
NASRDA	National Space Research and Development Agency
OIB	ocean island basalt
ppb	parts per billion ($1/10^9$)

ppm	parts per million ($1/10^6$)
SMOW	standard mean ocean water
ICP-OES	inductively coupled plasma optical emission spectrometry
∂	stable isotope ratio expressed relative to a standard

Author details

Uriah Lar[1]*, Isah Lekmang[1], Cedric Longpia[2] and Mohammed Tsalha[3]

1 Department of Geology, University of Jos, Jos, Nigeria

2 Department of Mining Engineering, Plateau State Polytechnique—Barakin Ladi, Jos, Nigeria

3 Centre for Geodesy and Geodynamics, Toro, Nigeria

*Address all correspondence to: ualexanderlar@yahoo.co.uk

References

[1] Grant NK, Rex DC, Freeth SJ. Potassium-argon ages and strontium isotope ratio measurement from volcanic rocks in northern Nigeria. Contributions to Mineralogy and Petrology. 1972;**35**(3):277-292

[2] Lar UA, Tsalha MS. Geochemical characteristics of the Jos-Plateau Basalts, North-Central Nigeria. Global Journal of Geological Sciences. 2005;**3**(2):4

[3] Akeh LE. Disaster threat in Nigeria. In: Proceedings of a Seminar for the 1999 World Disaster Prediction Campaign in Lagos, Nigeria; Vol. 5; 1999. pp. 56-86

[4] Black R, Girod M. Late Paleozoic to recent igneous activity in West African and its relationship to Basement structure. Journal of African Magmatism and Tectonics. In: Clifford TN, Gass IG, editors. Edinburgh: Oliver and Bovds. 1970:185-210

[5] Turner DC. Volcanoes of Biu basalt Plateau north-eastern Nigeria. Journal of Mining and Geology. 1977;**3**(6):49-62

[6] Macleod WN, Turner DC, Wright EP. The Geology of the Jos Plateau, vol. 1: General Geology. Geological Survey of Nigeria, Bulletin. 1971;**32**:91-95

[7] Patterson G. Lake Pidong—A Preliminary Survey of a Volcanic Crater Lake. UK: Dept. of Geography, University of Durham; 1986

[8] Lar UA, Lekmang IC, Shuaibu MT, Mahmud UM, Yakubu TA. Volcanoes of Nigerian: A preliminary study. Presented at the 7th Conference on African Geology, Assiut, Egypt; 2013

[9] Ugodulunwa FXO, Schoeneich K. Guide book on the field trip to Panyam volcanoes. In: 30th Annual Conference of Nigerian Mining & Geosciences Society; 1994

[10] Lar UA, Mohammed TS, Yakubu TA, Mahmud UM, Haruna ST, Samuel T. Development of geohazards map of Nigeria, mapping of volcanic edifices in Nigeria CGG. In: Toro Project; 2007. Unpublished Manuscript

[11] Irvine TN, Baragar WRA. A guide to the chemical classification of the common volcanic rocks. Canadian Journal of Earth Sciences. 1971;**8**(5): 523-548

[12] Rollinson HR. Using Geochemical Data: Evaluation, Presentation and Interpretation. UK: Longman Group Ltd.; 1993. 352p

[13] Cox KG, Bell JD, Pankhurst RJ. The Interpretation of Igneous Rocks. London: Allen and Unwin; 1979. 450p

[14] Pearce JA. A User's Guide to Basalts Discrimination Diagrams: Trace Elements Geochemistry of Volcanic Rocks. Geological Associations of Canada Short Course Notes. Vol. 121996. pp. 79-113

[15] Sun SS, McDonough WF. Chemical and isotopic systematics of oceanic basalts: Implications for mantle composition and processes. Journal of the Geological Society, London, Special Publication. 1989;**42**:313-345

[16] Langmuir CH, Hanson GN. An evaluation of major element heterogeneity in the mantle sources of basalts. Philosophical Transactions. Royal Society of London. 1980;**A297**:383-407

[17] Nakamura N. Determination of REE, Ba, Fe, Mg, Na and K in carbonaceous and ordiary chondrites. Geochimica et Cosmochimica Acta. 1974;**38**:757-775

[18] Sun SS, Nesbelt RW, Sharaskin AY. Chemical characteristics of mid-oceanic

ridge basalts. Earth and Planetary Science Letters. 1980;**44**:119-138

[19] Thompson RN, Morrison MA, Hendry GL, Parry SJ. British tertiary volcanic province. Scottish Journal of Geology. 1982;**18**:49-107

[20] Bonyton WV. Cosmochemistry of the Rare Earth Elements. Meteorite Studies. Amsterdam: Elsevier Sci. Publ. Co.; 1984. pp. 63-114

Chapter 3

P-Wave Teleseismic Tomography: Evidence of Imprints of Deccan Mantle Plume below the Kachchh Rift Zone, Gujarat, India

Prantik Mandal

Abstract

The Indian plate had experienced the Deccan volcanism at 65 Ma when it moved over the Re-union hotspot, which has altered lithospheric structure below the Kachchh rift zone (KRZ). To quantify the influence of Deccan volcanism on the crust-mantle, the present chapter focuses on the delineation of the upper mantle structure below the KRZ, through the modeling of crust corrected P-residuals and P-wave teleseismic tomography. The crust corrected normalized P-residuals suggest dominant negative residuals associated with the central KRZ, indicating crustal and lithospheric thinning below the KRZ. A low velocity down to a depth of 170 km below the central KRZ is detected through the teleseismic tomography using these P-residuals. However, these residuals also show positive values for the surrounding un-rifted zones. Note that a low shear velocity zone extending from 100–120 km to 170–220 km depth beneath the central KRZ has already been revealed by the modeling of P-RFs. This reduction in seismic velocity in the upper mantle could be explained by the presence of trapped carbonatite/partial melts related to the Deccan volcanism. The influx of volatile CO_2 emanating from the carbonatite melts in the asthenosphere might be generating lower crustal earthquakes occurring in the KRZ.

Keywords: teleseismic P-residuals, tomography, lithosphere, asthenosphere, Kachchh rift

1. Introduction

At ~860 Ma, Indian plate was broken from Antarctica and then it moved along with African plate until 184 Ma [1, 2]. Subsequently, it moved toward North until it got separated from Madagascar at 88 Ma [3]. At 65 Ma, it moved over the Reunion hotspot, which led to outpouring of Deccan volcanism resulting in the Deccan volcanic province occupying an area of 0.5 million km^2 with 2000 m thick basaltic lava in the western and central parts of the Indian subcontinent within a short duration of time [4].

Occurrence of the initial outpouring of Deccan volcanism in Pakistan (at ~72–73 Ma) has been evidenced by the available ^{40}Ar-^{39}Ar age data of the Reunion Island like alkali lavas [3]. The presence of relatively younger basaltic intrusions (~68.5 Ma) in the northern India [5] has supported the fact that India moved toward south. Also, younger basalts have been found further south [6]. Finally, at 65 Ma, the 95% basaltic outpouring of the Deccan volcanism took place in the western and central India [7]. This Deccan plume model has got further support from the presence of a low-velocity anomaly in the upper mantle extending up to a depth of 600 km below the north-western India as modeled by the regional mantle P-wave seismic tomography [8]. The presence of high $^3He/^4He$ ratio in Rajasthan has provided further support for the Deccan plume model [5]. Adding credence to this plume theory, local earthquake tomography, surface wave dispersion and receiver function modeling studies have imaged crustal mafic underplating, Moho upwarping and asthenospheric thinning underlying the 2001 Bhuj earthquake region [9, 10]. Recently, the shear-wave splitting study suggested that the upper mantle anisotropy in the KRZ is contributed both by lithospheric frozen anisotropy and asthenospheric flow induced anisotropy, which could be inherited from the plume-lithosphere interaction during the Deccan/Reunion plume episode (~65 Ma) [11]. These observations suggest that imprints of the Deccan/Reunion mantle plume are still present in the crust and upper mantle below the north-western region.

The Kachchh region has been experiencing earthquakes since historical times [12]. The region has already experienced seven M6 earthquakes including two M_w 7.7 earthquakes in 1989 and 2001. The latter event has claimed a death toll of 20,000 people. The aftershock activity of this 2001 earthquake is continuing until today, with regular occurrences of M_w 3 events and occasional occurrences of M_w 4 events. The aftershock activity of the 2001 Bhuj earthquake is still continuing that includes 15 $M_w \geq 5$, about 300 $M_w \geq 4$ and about 6000 $M_w \geq 3$ events. We feel that the enigmatic seismicity associated with the Kachchh rift zone is linked with its above-discussed unique Geodynamic history. Aiming at understanding the influence of crustal-mantle structure in the genesis of uninterrupted occurrences of earthquakes since the occurrence of the 26 January M_w 7.7 Bhuj earthquake, in this chapter, the crust corrected P-wave residuals are estimated at 14 broadband stations and a 3-dimensional P-wave teleseismic tomography is performed using the estimated crust corrected residuals. Finally, modeling results are interpreted concerning the geodynamical processes responsible for generating intraplate earthquakes occurring in the Kachchh region.

2. Seismic network and data

For the present study, broadband digital waveforms of 241 teleseismic events from 14 three-component broadband stations in Kachchh, Gujarat are used (**Figure 1a** and **b**). The sampling rate of recording is 50 sps. Station spacing in the above seismic network is 30–100 km, however, for our study we use data from 20 broadband stations consisting of 14 NGRI and 6 ISR stations as shown by a square in **Figure 2**. The station spacing for the network consisting of these 20 stations is 30–60 km. We selected 241 teleseismic events with magnitude ranging from 5.9 to 8.2, with epicentral distances of 29–90° (**Figure 1b**) with reference to the center of the network.

First, each trace of an event is used to pick the arrival time of the first P-wave maximum amplitude (either peak or trough). Then, these picks are correlated within the network. Following this procedure, the first P-wave onset times are picked from selected highest-quality traces recorded at different stations for one event. Here, the uncertainty of the picking is used to decide a quality factor for each measurement. Note that most of the measurements are found to be having an uncertainty of ±0.05 s. Finally, these quality factors are used to estimate the average data error, which is found to be ±0.06 s.

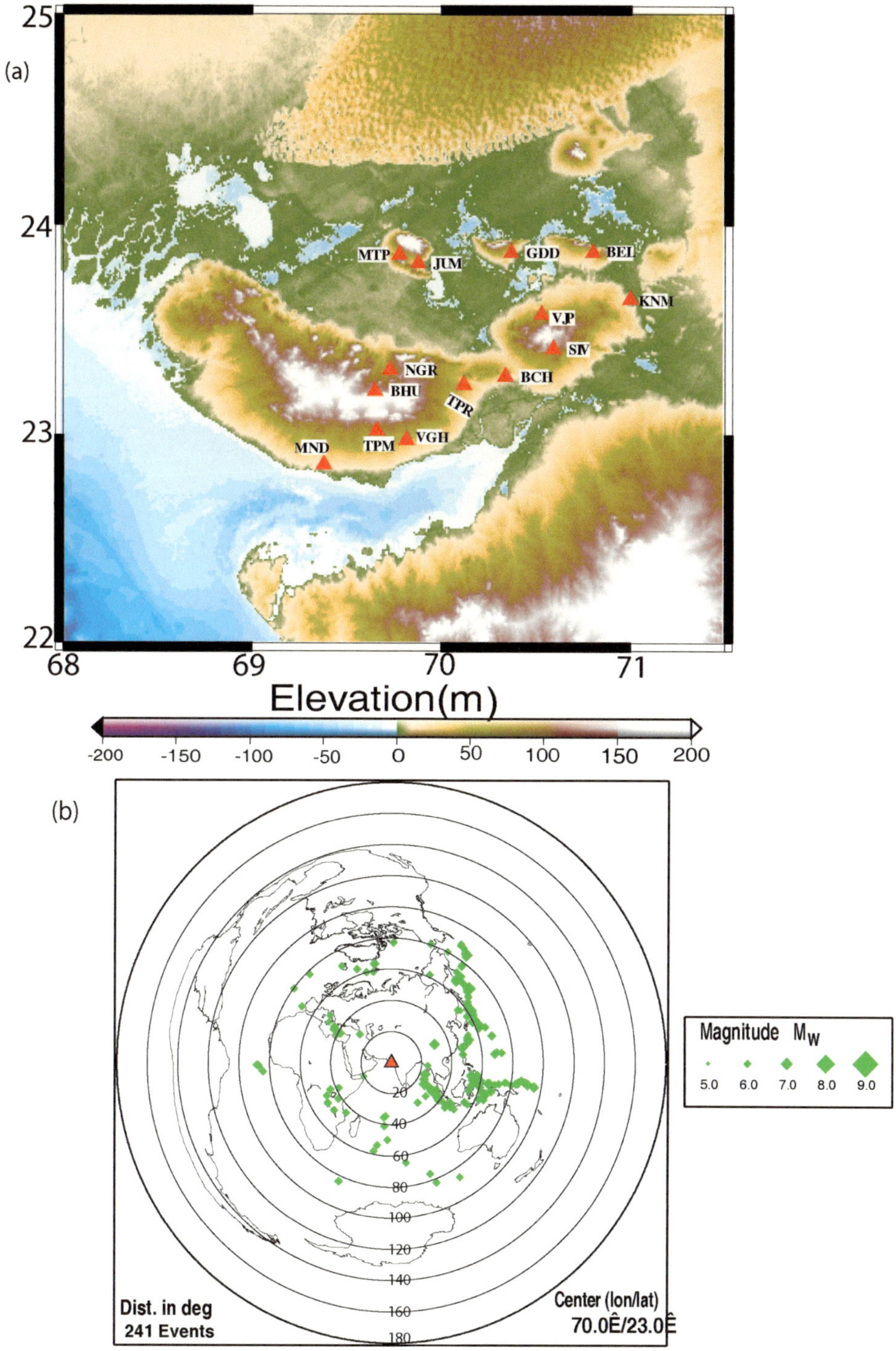

Figure 1.
(a) Elevation (in m) map showing station distribution in Kachchh, Gujarat. Filled red triangles mark the broadband seismograph stations (SIV—Sivlok; VJP—Vajepar; KNM—Kanmer; BCH—Bhachau; BEL—Bela; GDD—Gadhada; JUM—Jumkunaria; MTP—Motapaya; NGR—Nagor; BHU—Bhuj; NPR—Narayanpar; TPM—Tapar Mundra; MND—Mandvi; VGH—Vaghura; TPR—Tapar Anjar; and NDD—New Dudhai). The inset shows the key map for the area, where the study area is shown by a red open square. K and KL mark Kachchh, Gujarat and Killari, Maharashtra, respectively. An arrow shows the location of Cambay. The black filled portion marks the areal extent of Deccan volcanic province (DVP) in India. (b) Epicentral plot of 241 teleseismic events of M_w 6.0–8.4, whose broadband data are used for our P-wave teleseismic tomography study. A red triangle and green diamond symbols mark the center of our network (lat. 70°, long. 23°) and epicenters of selected teleseismic events. The size of the diamond symbols vary depending on their sizes.

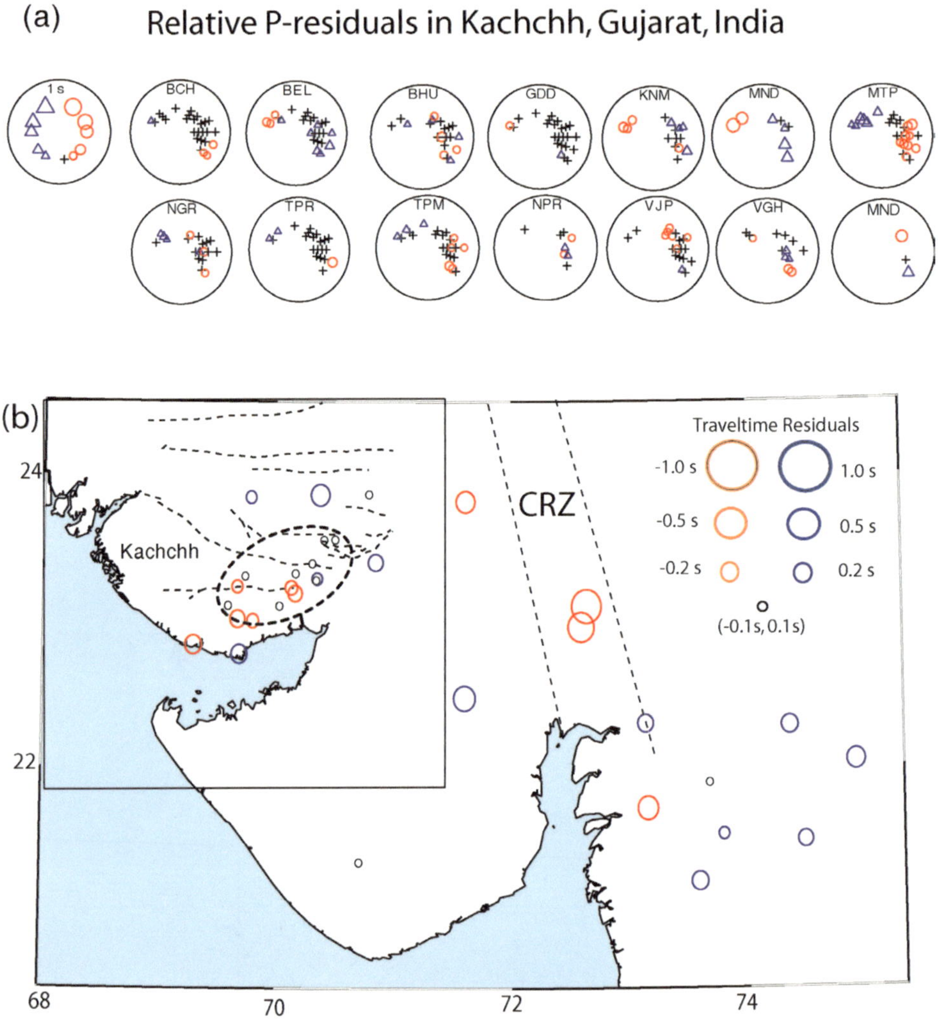

Figure 2

(a) Crust corrected relative P-residuals in Gujarat. (b) Directional mean of P-residuals by subtracting an average residual from the eight stations in Kachchh. Red open circles mark negative residuals while blue open circles show positive residuals. CRZ marks the Cambay rift zone, which is shown by dotted lines. Black dotted elliptical area mark the central Kachchh rift zone characterized by negative residuals. And, black square area shows our study area considered for our P-wave teleseismic tomography.

3. Calculation of crust-corrected P-relative residual

For picking P, the measured waveforms from our network are bandpass filtered using the inbuilt WWSSN filter of the seismic handler. This narrow frequency band allows us to study the uppermost mantle without caring about finite frequency effects [13]. In order to calculate relative residuals, only first strong and clear phase (peak or trough) are picked from different stations, with the help of high magnification on the computer screen. Since the sampling rate used for data recording is 50 sps, thus, phases could be picked manually with a precision of at least 0.08 s. Thus, the most of the estimated relative residuals are found to have an uncertainty of less than 0.1 s.

The analysis of broadband waveforms of teleseismic events are used here to estimate travel time residuals, as well as deviations in slowness and back azimuth about the standard earth model iasp91 [14]. A travel time residual can be defined as

$$r_{ij} = t_{ij}^{m} - t_{ij}^{t} \quad (1)$$

where t_{ij}^{m} is the measured arrival time of a teleseismic phase j from an event at ith station and t_{ij}^{t} is the theoretical arrival time derived from an earth model (e.g., iasp91). Now, a systematic bias to r_{ij} could be resulted from following sources:

i. the uncertainties in origin time and hypocenter location of the teleseismic events

ii. the source side heterogeneity.

To isolate information underneath the receiver network, the average of all measurements (n) of a phase j is subtracted from the estimated relative residual using the following equation,

$$r_{ij}^{r} = r_{ij} - (1/n)\sum_{i=1}^{n} rij \quad (2)$$

Thus, the residual part, which is common in all measurements, is eliminated. Therefore, the remaining part of traveltime residual, which is caused by heterogeneity below the receivers, defines the relative residual. This normalization (Eq. (2)) produces an equal amount of positive and negative residual times per phase with a zero mean. Now, to use these relative residuals in tomographic inversions, a weighting scheme following Evans and Achauer [15] has been used here. The weights w_{ij} are assigned while picking the arrival times based on arrival time uncertainties (d) and SNR of the picks. We use four different weights (1, 2, 3, and 4): 1 = 1.0 for $d \le 0.05$ s, 2 = 0.2 for $0.05 \text{ s} < d \le 0.1$ s, 3 = 0.1 for $0.1 \text{ s} < d \le 0.2$ s and 4 for $d \le 0.2$ s. The picks with quality 4 are excluded from the further analysis.

3.1 Estimation of relative travel-time residual

After estimating relative travel time residuals using Eq. (2), these residuals are corrected from already known heterogeneous crustal anomalies and the travel time anomalies from the top sediments, which generally cause unwanted smearing effects of the residuals along the steep teleseismic ray paths [15, 16]. An available detailed crustal velocity model is used for this purpose and then these residuals are corrected from crustal travel time anomalies (r_{ij}^{crust}) relative to the background earth model using the relation:

$$t_{ij,c}^{m} = t_{ij}^{m} - r_{ij}^{crust} \quad (3)$$

3.2 Crustal correction

It is well-established fact that significant travel time residuals can be caused by the heterogeneous continental crust. Further, large delays of the order of 1 s can be resulted from the low-velocity sediments in the rift basins (like Kachchh) or/either the Moho topography, which laterally replaces mantle (Vp ~ 8 km/s) and crustal (Vp ~ 6.5 km/s) material. Thus, the relative residuals should be corrected from such known crustal residuals to study the deep mantle structure. Therefore, these crustal anomalies should be determined a priori and subtracted [16, 17]. Following Martin et al.'s [16] approach, here the crustal corrections are applied using the iasp91 reference earth model consisting of a 20 km thick upper crust with a Vp of 5.8 km/s and a 15 km thick lower crust with a Vp of 6.5 km/s. To apply crustal correction, top sedimentary layer is assumed to be composed of Cenozoic sediments (Vpsed ~ 2.41–4.60 km/s) and Jurassic/Mesozoic sediments (Vpsed ~ 4.61–5.58 km/s). The estimated corrections of the travel time residuals for such 1-D known crustal velocity models in the region vary from 0.29 to 0.79 s for P-waves.

A total of 241 teleseismic events (M_w 6.0–8.0) have been analyzed to estimate crust corrected P-residuals using Eq. (3), they are found to be stable for stations in the Kachchh network (**Figure 2**). From **Figure 2**, we notice that there is no clear anisotropic signal in the P-residuals in most of Kachchh region except the Motapaya (MTP) station, which is the northernmost station in Kachchh. **Figure 2** shows a plot of estimated crust corrected relative P-residuals at 14 stations as a function of slowness (p) (0–10 s/o along the radial axis) and back azimuth (BAZ 0–360°) to the event. In general, early P-arrivals are noticed for events in the east of the network while delayed P-arrivals are observed for events in the west. Most of the stations, which are lying in the Deccan basalt covered regions in the Kachchh and Cambay rift zones, show negative residuals. This could be attributed to the presence of significant crustal and lithospheric upwarping below these rift zones [18, 19]. At MTP, travel time residuals show images of the spatial variation of wave propagation anomalies of P-waves, which can be attributed to an anisotropy structure of the mantle domains (**Figure 2a**). This anisotropic nature of P-waves is not seen for other stations due to fewer numbers of observations.

The directional term of relative crust-corrected P-residuals is calculated by dividing the 0–360 to 18 quadrants, and then we calculate the average residuals of stations falling in a particular quadrant. And finally, we estimate the arithmetic mean of the average residuals of stations falling in all 18 quadrants. Subsequently, this mean of residuals is being subtracted from the relative crust-corrected P-residuals for different stations for estimating the directional term of relative crust-corrected P-residuals. Following the above procedure, we also computed directional term of relative crust-corrected P-residuals, which are plotted in **Figure 2b**. The directional mean of P-residuals, which are obtained by subtracting average residuals of all stations in Kachchh, is shown in **Figure 2b**. While **Figure 2b** represents the directional mean of P-residuals, which are calculated by subtracting average residual of 8 stations in Kachchh. Most interestingly, two distinct zones of negative P-residuals associated with Kachchh and Cambay rift zones are noticed from **Figure 2b**. These zones of negative P-residual probably mark the regions of marked lithospheric thinning. This model gets further support from available estimates of crustal and lithospheric thicknesses, suggesting a marked crustal and lithospheric thinning below the Kachchh and Cambay rift zones [18, 19].

4. P-wave teleseismic tomography

The teleseismic tomography code developed by Weiland et al. [20] and later modified by various investigators [21–24] has been used here to estimate the 3-D P-velocity structure down to 250 km underlying the Kachchh rift zone (KRZ). The data collected by the digital networks of NGRI (Hyderabad, India) and ISR (Gujarat, India) are combined, resulting in a dataset from a total of 59 stations (**Figure 2b**). Here, the P-wave teleseismic tomography is performed using relative crust-corrected P-residuals as estimated above. First velocity perturbations in this method are calculated in the orthogonal net of nodes approximating the volume under consideration. And, the trilinear interpolation is used to calculate velocity at adjacent nodes [25]. The initial velocity model of the upper mantle and theoretical travel times are set according to a reference Earth model IASP91 [14]. We run the tomographic code with three-dimensional ray tracing implemented by the Simplex method [25]. The kernel matrix is inverted by Singular Value Decomposition (SVD). The basic equation of the inversion of the tomographic code (i.e., TELINV) can be written as:

$$\mathbf{m}_{est} = \left(\mathbf{G}^T \mathbf{W}_D \mathbf{G} + \varepsilon^2 \mathbf{I}\right)^{-1} \mathbf{G}^T \mathrm{W}_D \mathbf{d} \tag{4}$$

where $\mathbf{mes}_t$ and **G** are estimated model parameters (i.e., velocity perturbations) and matrix of partial derivatives with respect to the model parameters, respectively. $\mathbf{W}_D$ is weighting matrix of data, where weights are set according to quality factors assigned to individual arrival time picks. ε^2 and **I** are a damping factor and a unit diagonal matrix, respectively. And, **d** is data vector (i.e., relative residuals) [26]. Several iterative cycles are performed for satisfying assumptions behind the linearization of the inversion. New ray-paths inside of the area studied are traced at each cycle using the improved velocity model retrieved in the previous step. Here, four iteration cycles are applied to reach such data variance which does not decrease with further iterations noticeably and stay above the twice the average data error.

4.1 Model parameterization

The primary study area extends about 320 km in E-W and 240 km in N-S, which covers the E-W trending Kachchh rift zone (**Figures 1a** and **2b**). Here, the volume below the central study area is approximated by the 3D (i.e., x, y and z directions) grid of nodes (17 × 18 × 15). For teleseismic tomography, a much large area covering 10,000 km × 10,000 km × 530 km, is assumed to stabilize the velocity perturbation outside the main study region. Seventeen X-nodes are distributed at distances of −5000 −240 −200 −160 −120 −80 −40 0 40 80 120 160 200 520 560 600 5000 km while eighteen Y-nodes are located at distances of −5000 −360 −320 −280 −240 −200 −160 −120 −80 −40 0 40 80 120 160 200 240 5000 km. And, fifteen nodes in Z directions are at depths of −5 10 70 50 90 130 170 210 250 290 330 370 410 450 490 km. The center of the array is assumed to be at latitude 23.3° and longitude 70.3°. Here, the teleseismic tomographic inversion is performed using a total of 1788 P-residuals, which are estimated using vertical component of broadband seismograms of 241 good teleseismic events (with epicentral distances between 30° and 90°) recorded at 59 stations. Here, the travel time residuals range from −2.276 s to 2.096 s. This study results in an average data error of 0.062 s. For the present study, a travel-time residual of 2.5 s is used while a ε^2 damping factor of 100 is used for the inversion. To minimize effects of potential inaccuracies of the 3D crustal model applied in the inversion, the first inverted layer of nodes is assumed at 50 km depth. Several different vertical parameterizations with irregular and smaller spacing are being tested, but the variance reduction, as well as the diagonal terms of the resolution matrix, decreased rapidly. In total, the inversion is performed for 980 model parameters.

The crustal structure cannot be resolved through teleseismic tomography due to the sub-vertical directions of the incoming rays. However, the heterogeneous, complex crustal structure can affect the inverted travel-times significantly [27–29]. Further, the size of crustal heterogeneities is often exceeded by the spacing between stations. Therefore, crustal corrections are inevitable to apply on the travel-time residuals, before inverting the residuals for the upper mantle velocity structure. Here, following the standard procedure [17, 29] crustal corrections as discussed above are applied to the dataset. Modeled dVp (%) tomograms at different depths are shown in **Figure 3a–d**.

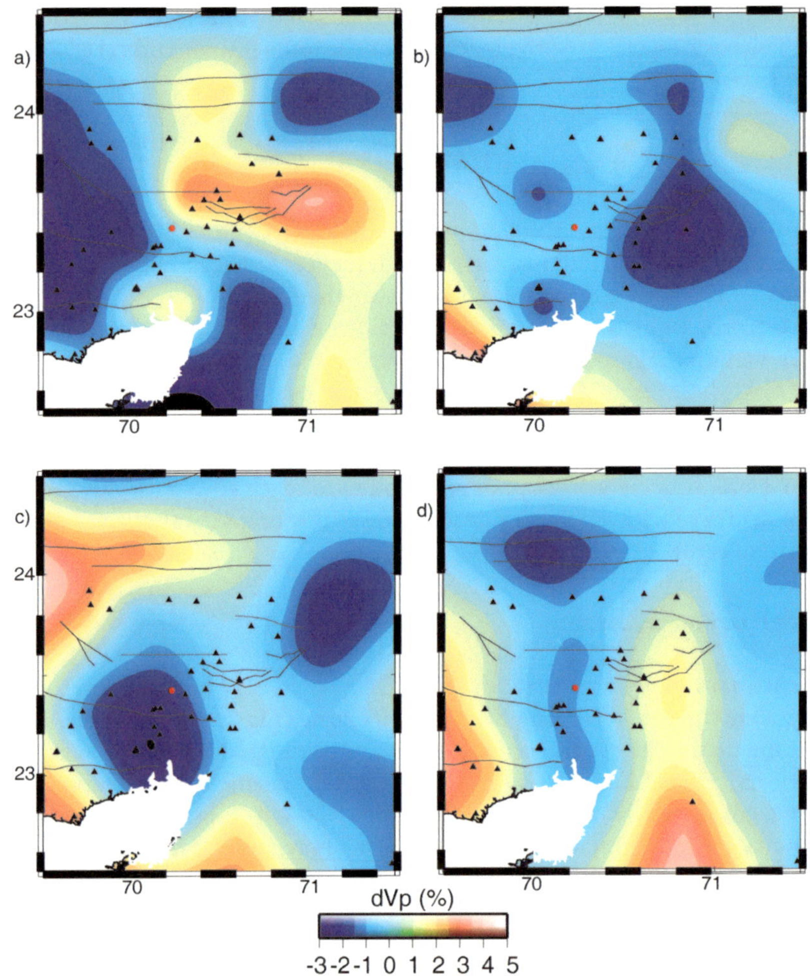

Figure 3
P-wave velocity perturbations at horizontal slices imaged at (a) 50, (b) 90, (c) 130, and (d) 170 km depths, by the P-wave tomography using the crust corrected relative P-residuals. Major faults (solid white lines): ABF—Allah bund fault; IBF—Island belt fault; KMF—Kachchh mainland fault; KHF—Katrol hill fault; NPF—Nagar Parkar fault; BF—Banni fault. And, NWF (north Wagad fault), the causative fault for 2001 Bhuj earthquake and Gedi fault, are shown by dotted white lines. Locations of mobile broadband stations are marked by gray open triangles. A black dotted elliptical area mark the central KRZ.

4.2. Resolution analyses

Resolution analyses of tomographic models are essential for distinguishing real structures from artifacts caused by methods used and for identifying well-resolved model parameters. The reliability of tomographic images have been analyzed by performing a checkerboard test, which is performed with the ray geometry of the Kachchh network, with 3D ray-tracing, and with the damping factor of 100.

To verify sensitivity in the whole volume studied, a checkerboard test is performed [30]. For this, a net of alternating anomalies of +5% and −5% in nodes is constructed at depths of 50, 90, 130, and 170 (**Figure 4a–d**), leaving the layers in between them unperturbed as well as in the remaining parts of the model [30]. The input anomalies are recovered well at depths ranging from 50 to 170 km (**Figure 4e–h**). The low-velocity anomalies in the central part are not resolved successfully at ~210 km depth, which could be attributed to the well-known vertical smearing dominated in the inversions of teleseismic data.

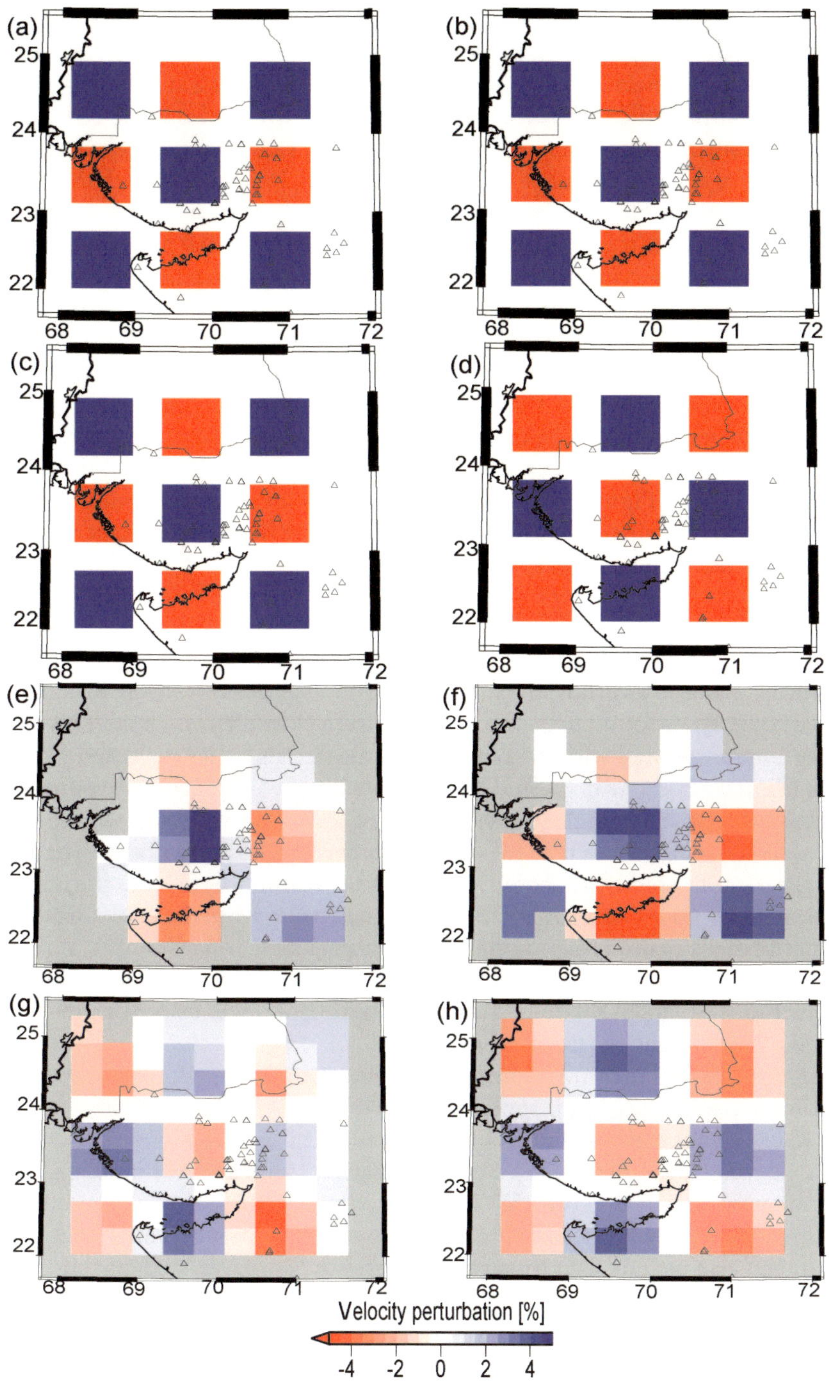

Figure 4
Input 3-D synthetic P-wave velocity model with alternative grids of high (±5%) and low (−5%) velocity for the checkerboard test at (a) 50, (b) 90, (c) 130, and (d) 170 km depths. Results of checkerboard test showing retrieval of synthetic velocity perturbations at (e) 50, (f) 90, (g) 130, and (h) 170 km depths.

5. Results and discussions

Large negative travel time residuals associated with the central Kachchh rift zone are detected (marked by a black dotted elliptical area shown in **Figure 2b**), suggesting relatively thin lithosphere underlying the central KRZ while surrounding unrifted parts show positive travel time residuals suggesting relatively thick lithosphere as also observed in the estimates of lithospheric thicknesses [18]. Large negative travel time residuals are also found in the Cambay rift zone (CRZ) (**Figure 2b**) suggesting relatively thin lithosphere underlying the CRZ as also modeled by an S-RF imaging and surface wave dispersion studies [19, 31]. Note that the estimated directional means of relative P-residuals show a good spatial correlation with the major tectonic regional features. The most negative values (though with large standard deviations) follow the track of the Cambay rift or Deccan plume and basaltic region of Kachchh. This observation shows the robustness of our modeling results.

The horizontal slices of dVp tomograms at different depths (50, 90, 130, and 170 km) suggest that a prominent low-velocity zone (~−2 to −4%) is present below the central Kachchh rift zone extending from 50 to 170 km depths (**Figure 3a–d**). At 50 km depth, a high-velocity structure is also noticed in the region north-east of the 2001 mainshock location (**Figure 3a**), which vanishes at deeper depths. But, the low-velocity zone in the KRZ becomes a bit subdued (~−1 to −3%) between 150 and 170 km depths (**Figure 3c** and **d**). However, this low-velocity zone also exists at 210 km depth, which might be an artifact due to the poor sampling of rays for the deeper zone. Tomograms also clearly bring out the fact that in the upper mantle (at 70–170 km depth) the P-wave velocity is found to be higher in the western part in comparison to that on the eastern part of the Kachchh region. In the shallower depths (at 50 km), tomogram suggests that the central KRZ, which is characterized by the crustal as well as lithospheric thinning and negative traveltime residuals (**Figure 2b**; [18]), is found to be associated with the low dVp anomaly. This low-velocity zone associated with the central KRZ also exists between 90 and 170 km depths (**Figure 3b–d**), which could be attributed to the presence of carbonatite/partial melts related to the Deccan volcanism of 65 Ma [19, 31]. Note that the presence of carbonatite melts in the shallow upper mantle depths below the KRZ has also been proposed by an isotopic ratio study of xenoliths [7] and the modeling of SKS splitting [11]. Thus, the volatile CO_2 (emanating from the crystallization of carbonatite melts in the asthenosphere), can reach lower crustal depths, through deep-seated faults, which might be facilitating the deeper circulation of metamorphic fluids/volatile CO_2, thereby, the generation of lower crustal earthquakes occurring in the Kachchh rift zone [18].

The most robust outcome of the tomography, which does not vary with regularization or parameterization applied, is that low-velocity perturbations prevail in the Kachchh rift zone (as shown by dotted elliptical area in **Figure 2b**) down to ~170 km depth (see **Figure 3a–d**). Similarly, earlier studies [8, 10] indicated that the upper mantle beneath the entire KRZ is characterized by lower velocities, about the surrounding area down to these depths. These studies do not exploit only teleseismic data but also include data from regional events. Moreover, regional and global teleseismic tomography studies [8, 32] show the upper mantle beneath the Kachchh region as a part of an extensive low-velocity heterogeneity located in western India, extending down to about 700 km. The regional teleseismic tomography from the western India indicates relatively small velocity variations (±1%), both in size and in amplitude. Such details are below resolution level of global or regional studies which invert data of permanent observatories only. Thus, our results can be viewed as being in agreement with these larger-scale investigations.

6. Conclusions

The variation of crust corrected travel time residuals correlate well with our estimates of crustal and lithospheric thicknesses in the study region. The large negative residuals are found to be associated with the Kachchh and Cambay rift zones, which are characterized by a marked crustal and asthenospheric thinning while positive residuals describe the unrifted regions.

Our tomograms reveal a distinct low dVp anomaly at 50–170 km depths underlying the central Kachchh rift zone. However, this anomaly vanishes for deeper regions (>210 km) due to poor ray sampling. At 90–130 km depths, the resulting tomographic model reveals a slower P-wave speed on the eastern part of Kachchh compared to that of the western part. A slow dVp anomaly is mapped at 50–170 km depths below the central KRZ, which is also characterized by the crustal as well as lithospheric thinning and negative traveltime residuals. This upper mantle low dVp anomaly could be attributed to the presence of carbonatite/partial melts related to the Deccan volcanism of 65 Ma. We propose that the volatile CO_2, which is emanating from the crystallization of carbonatite melts in the asthenosphere, is playing a crucial role in generating lower crustal earthquakes occurring in the Kachchh rift zone [18].

The checkerboard test results suggest that our tomographic model is quite well resolved for the central KRZ at 50–170 km depths while tomographic model at more than 210 km depth is poorly resolved due to poor ray sampling. The checkerboard test reveals that 85–90% of the synthetic velocity anomaly could be retrieved at 50–170 km depths below the central KRZ (**Figure 4e–h**). Thus, the velocity model below the KRZ seems to be well-resolved. However, the retrieval of synthetic velocity for the surrounding regions are not very good (<80%). To increase the resolution of the tomographic images of the upper mantle velocities, a future study could be planned to combine Kachchh dataset with those from other broadband stations in Gujarat and Rajasthan, using more teleseismic events with epicentral distances between 30 and 90° and back-azimuths between 180 and 270°.

Acknowledgements

The author is grateful to the Director, Council of Scientific and Industrial Research—National Geophysical Research Institute (CSIR-NGRI), Hyderabad, India, for his support and permission to publish this work. This study was supported by the Indo-Czech (CSIR-ASCR) collaborative project no, IND 2012/19.

Author details

Prantik Mandal
CSIR-NGRI, Hyderabad, India

*Address all correspondence to: prantikmandal62@gmail.com

References

[1] Fitzsimons ICW. Proterozoic basement provinces of southern and southwestern Australia, and their correlation with Antarctica. In: Yoshida M, Windely BF, Dasgupta S, editors. Proterozoic East Gondwana: Supercontinent Assembly and Breakup. Vol. 206. London: Geological Society; 2003. pp. 93-130 (Special Publication)

[2] Maji AK, Goon S, Bhattacharya A, Mishra B, Mahato S, Bernhardt HJ. Proterozoic polyphase metamorphism in the Chhotanagpur gneissic complex (India), and implication for trans-continental Gondwanaland correlation. Precambrian Research. 2008, 2008;**162**:385-402

[3] Mahoney JJ, Duncan RA, Khan W, Gnos E, McCormick GR. Cretaceous volcanic rocks of the south Tethyan suture zone, Pakistan: Implications for the Réunion hotspot and Deccan traps. Earth and Planetary Science Letters. 2002;**203**:295-310

[4] Courtillot V, Besse J, Vandamme D, Montigny R, Jaeger JJ, Cappetta H. Deccan flood basalts at the cretaceous/tertiary boundary? Earth and Planetary Science Letters. 1986;**80**:361-374

[5] Basu AR, Renne PR, Dasgupta DK, Teichmann F, Poreda RJ. Early and late alkali igneous pulses and a high-3He plume origin for the Deccan flood basalts. Science. 1993;**261**:902-906

[6] Baksi AK. Critical evaluation of the age of the Deccan traps, India: Implications for flood-basalt volcanism and faunal extinctions. Geology. 1987;**15**:147-150

[7] Sen G, Bizimis M, Das R, Paul DK, Ray A, Biswas S. Deccan plume, lithospheric rifting, and volcanism in Kutch, India. Earth and Planetary Science Letters. 2009;**277**:101-111

[8] Kennett BLN, Widiyantoro S. A low seismic wavespeed anomaly beneath northwestern India: A seismic signature of the Deccan plume? Earth and Planetary Science Letters. 1999;**165**:145-155

[9] Mandal P, Pujol J. Seismic imaging of the aftershock zone of the 2001 Mw 7.7 Bhuj earthquake, India. Geophysical Research Letters. 2006;**33**:L05309(1-4)

[10] Mandal P. Passive source seismic imaging of the crust and upper mantle beneath the 2001 Mw 7.7 Bhuj earthquake region, Gujarat, India. Bulletin of Seismological Society of America. 2012;**102**(1):252-266

[11] Mandal P. Upper mantle seismic anisotropy in the intra-continental Kachchh rift zone, Gujarat, India. Tectonophysics. 2011;**2011**(509):81-92

[12] Rajendran CP, Rajendran K, Thakkar M, Goyal B. Assessing the previous activity at the source zone of the 2001 Bhuj earthquake based on the nearsource and distant paleoseismological indicators. Journal Geophysical Research. 2008;**113**:B05311-1-B05311-17

[13] Nolet G, Dahlen FA. Wave front healing and the evolution of seismic delay times. Journal of Geophysical Research. 2000;**105**:19043-19054

[14] Kennett BLN, Engdahl ER. Travel times for global earthquake location and phase identification. Geophysical Journal International. 1991;**105**:429-465

[15] Evans JR, Achauer U. Teleseismic velocity tomography using the ACH method: Theory and application to continental-scale studies. In: Iyer HM, Hirahara K, editors. Seismic Tomography: Theory and Practice. London: Chapman & Hall; 1993. pp. 319-360

[16] Martin M, Ritter R, the CALIXTO working group. High-resolution teleseismic body-wave tomography beneath SE Romania-I. Implications for three-dimensional versus one-dimensional crustal correction strategies with a new crustal velocity model. Geophysical Journal International. 2005;**162**:448-460

[17] Lippitsch R, Kissling E, Ansorage J. Upper mantle structure beneath the alpine orogen from high resolution teleseismic tomography. Journal of Geophysical Research. 2003;**108**:2376. DOI: 10.1029/2002JB002016

[18] Mandal P. Influence of Deccan Volcanism and Synrift magmatism on the crust-mantle structure and its implications toward the seismogenesis of earthquakes occurring in the Kachchh rift zone: In a special volume of "Tectonics of the Deccan Large Igneous Province". In: Mukherjee S, Misra AA, Calvès G, Nemčok M, editors. Tectonics of the Deccan Large Igneous Province. Vol. 445. London: Geological Society; 2017. pp. 189-218 (Special Publications)

[19] Kumar P, Sen G, Mandal P, Sen MK. Shallow lithosphere-asthenosphere boundary beneath Cambay rift zone of India: Inferred presence of carbonated partial melt. Journal of Geological Society of India. 2016;**88**:401-406

[20] Weiland CM, Steck LK, Dawson PB, Korneev V. Nonlinear teleseismic tomography at long Vally caldera, using three dimensional travel time ray tracing. Journal of Geophysical Research. 1995;**100**:20379-20390

[21] Achauer U. A study of the Kenya rift using delay-time tomography analysis and gravity modelling. Tectonophysics. 1992;**209**:197-207

[22] Kissling E, Spakman W. Interpretation of tomographic images of uppermost mantle structure: Examples from the western and Central Alps. Journal of Geodynamics. 1996;**21**:97-111

[23] Shomali ZH, Roberts RG, Pedersen LB. The TOR working group: Lithospheric structure of the Tornquist zone resolved by nonlinear P and S teleseismic tomography along the TOR array. Tectonophysics. 2006;**416**:133-149

[24] Eken T, Shomali H, Roberts R, Bodvarsson R. Upper mantle structure of the Baltic shield below the Swedish National Seismological Networks (SNSN) resolved by teleseismic tomography. Geophysical Journal International. 2007;**169**:617-630

[25] Steck LK, Prothero WA. A 3-D ray-tracer for teleseismic body-wave arrival times. Bulletin of the Seismological Society of America. 1991;**81**:1332-1339

[26] Menke W. Geophysical Data Analysis: Discrete Inverse Theory. College of Oceanography. Oregon State University: Academic Press, Inc.; 1994. p. 239

[27] Arlitt R, Kissling E, Ansorge J. Three-dimensional crustal structure beneath the TOR array and effects on teleseismic wavefronts. Tectonophysics. 1999;**314**:309-319

[28] Waldhauser F, Lippitsch R, Kissling E, Ansorge J. High-resolution teleseismic tomography of the upper mantle structure using an aapriori three dimensional crustal model. Geophysical Journal International. 2002;**150**:403-414

[29] Sandoval S, Kissling E, Ansorge J, SVEKALAPKO Seismic Tomography Working Group. High-resolution body wave tomography beneath the SVEKALAPKO array: I. A priory three dimensional crustal model and associated travel-time effects on teleseismic tomographic wave fronts. Geophysical Journal International. 2003;**153**:75-87

[30] Sandoval S, Kislling E, Ansorge J, SVEKALAKO Seismic Tomography working Group. High-resoution

body wave tomography beneath the SVEKALAPKO array: II. Anomalous upper mantle structure beneath the Central Baltic Shield. Geophysical Journal International. 2004;**157**:200-214

[31] Mandal P. Upper mantle shear velocity structure below northwestern India based on group velocity dispersion. Bulletin of Seismological Society of America. 2015;**105**:252-266

[32] Replumaz A, Negredo AM, Villaseñor A, Guillot S. Indian continental subduction and slab break-off during tertiary collision. Terra Nova. 2010;**22**:290-296. DOI: 10.1111/j.1365-3121.2010.00945.x

Chapter 4

Eruption Types and Processes in the Guamsan Caldera, Korea

Sang Koo Hwang

Abstract

The Guamsan caldera is associated with the Guamsan Tuff and rhyolitic intrusions. The Guamsan Tuff consists of dominant ash-flow tuffs with some volcanic breccias and fallout tuffs. The breccias comprise block and ash-flow breccia near a vent and caldera-collapse breccia near a ring fracture. The lower member of the ash-flow tuffs is produced from pyroclastic flow-forming eruptions with any ash-cloud falls on the flow units, whereas the upper member is formed by many ash-flow from boiling-over eruptions. The rhyolitic intrusions are divided into intracaldera plug and ring dikes. The volcanic activities in the caldera exhibit the volcanic processes along a caldera cycle together with eruption types during 63.77–60.1 Ma. The activities began with pelean eruption that occurred with block and ash-flows from lava dome collapse, progressed through expanded pyroclastic flows and ash-cloud falls by pyroclastic flow-forming eruptions from a single central vent, and transmitted with non-expanded ash-flows from boiling-over eruptions along multiple ring fissure vents. Then the caldera collapse induced any translations into multiple ring fissure vents from an earlier single central vent. The boiling-over eruptions were followed by effusive eruptions along which rhyolitic magma was injected as a small plug and ring dikes with some lava domes on the surface.

Keywords: Guamsan caldera, Guamsan tuff, ring dikes, pyroclastic flow-forming eruptions, boiling-over eruptions, effusive eruptions

1. Introduction

Juwangsan volcanic field, located in the northeastern part of Gyeongsang basin, comprises several calderas associated with the succession of ash-flow tuff—caldera—ring dike. The evolution of Guamsan caldera, in the central part of the volcanic field, has been reported on in detail by a previous study [1, 2]. However, the eruption types in the caldera were not investigated. This study thus intends to interpret eruptive types and volcanic processes of the volcanic rocks associated with the caldera.

The strata in the Gyeongsang basin are so deeply eroded that the volcanic rocks are dominantly left inside the calderas. Likewise, the Guamsan caldera area has been so significantly eroded and so deep in valleys that it provides us with an excellent laboratory field that not only reveals almost all of the volcanic rocks related to the caldera but also reveals the intrusions corresponding to volcanic roots. Thereby, the Guamsan caldera area exposes the Guamsan Tuff as well as diverse intracaldera

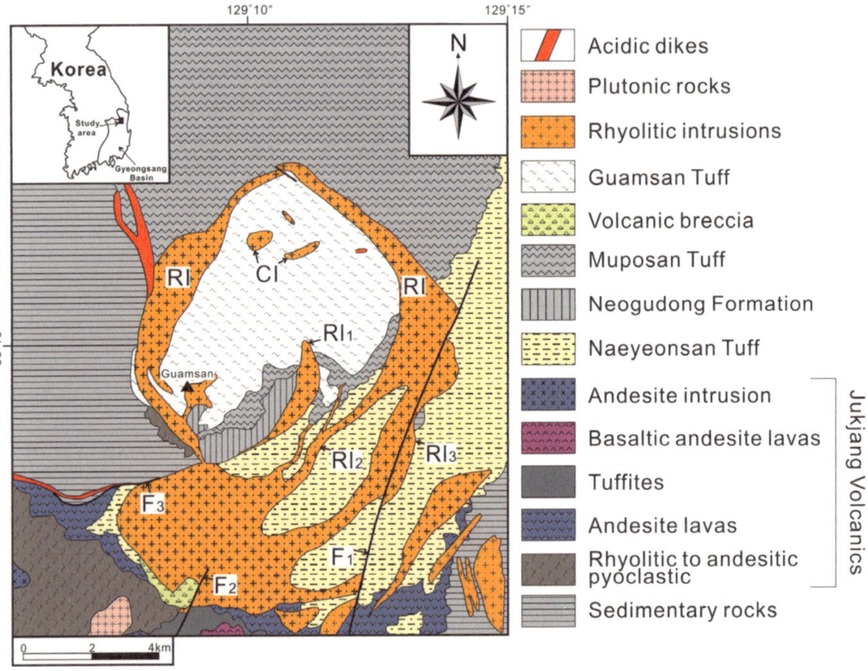

Figure 1.
Generalized geological map in and around the Guamsan caldera in the northeastern Gyeongsang basin. RI, ring intrusions (RI, inner ring dike; RI_2, intermediate ring dike; RI_3, outer ring dike); CI, intracaldera intrusions; F1, Sampo fault; F2, Jayangcheon fault; EW, trending fault.

intrusions and ring dikes (**Figure 1**). The lithofacies and sequences of these extrusive rocks and intrusive rocks are sufficient to interpret eruption types and reconstruct volcanic processes in the Guamsan caldera.

Eruption types and volcanic processes before and after caldera collapse will be discussed in this study. The ultimate results reveal that the volcanic activities in Gyeongsang Basin are not only related with caldera volcanisms but are also significant for understanding the characteristics of the igneous processes. Further, the consequences will contribute to the understanding of other volcanisms and processes in the calderas as well as their comparative effects.

2. Geological setting

The Gyeongsang basin has a broad distribution of volcanic rocks which are products of the Late Cretaceous to Early Paleogene calc-alkalic volcanism in the subduction zone along the Eurasian continental margin [3–6]. These volcanic rocks are mostly distributed in the Yucheon subbasin and are also found in the region between southeastern Yeongyang subbasin and mid-eastern Euiseong subbasin; this region belongs to the Juwangsan volcanic field.

The volcanic rocks occurring in the field mostly consist of extrusive rocks accompanied by small amounts of intrusive rocks. The extrusive rocks are placed on sedimentary rocks of the Hayang Group and can be roughly categorized into lower basic to intermediate volcanic rocks and upper acidic volcanic rocks. The former extrusive rocks comprise the stratigraphic units of Daejeonsa Basalt, Ipbong Andesite, and Jukjang Volcanics. The latter extrusive rocks consist of Jipum Volcanics (68.5 Ma in [7]), Juwangsan Tuff, Naeyeonsan Tuff, Neogudong Formation, Muposan Tuff (67.08 Ma in [8]), and Guamsan Tuff (63.77 ~ 60.1 Ma in [9]).

The intrusive rocks comprise rhyolitic intrusions relevant to the caldera as well as biotitic granite and felsite dikes irrelevant to the caldera. The rhyolitic intrusions correspond to the volcanic rocks of postcaldera, which can be divided into intracaldera intrusions and ring dikes. The biotitic granite is exposed as small stocks, whereas the felsite is exposed as linear dikes.

In the area, the geology is cut by the Sampo fault and Jayangcheon fault, the strike-slip fault running southwest from northeast and is depressed by a fault running west from east (**Figure 1**).

3. Guamsan caldera

The Guamsan caldera is bound along the structural line as determined by the outer ring dike. The caldera is approximately 9.2 km in maximum diameter and 8.0 km in minimum diameter [1]; the resulting internal area is then approximately 66.0 km in [2].

The intracaldera Guamsan Tuff has a contact with the underlying Muposan Tuff, with the ring dike intervening between two units, suggesting that the former unit has been subsided as compared to the latter unit. This can be counted as direct evidence of the subsidence by the collapse of a caldera. The welding foliation and bedding in the volcanic and sedimentary rocks generally represent a basin structure like a bowl shape, which has steep to gentle dips inwardly from the caldera margin. The structure suggests a direct subsidence from the collapse of the caldera. The caldera block shows an asymmetrical feature that was collapsed to 900 m along the northern margin, whereas it was collapsed to 300 m along the southern margin [1].

Therefore, the caldera was formed by down-sagging and ring-faulting. Based on comprehensive integration of these evidences, the caldera is classified as one of the asymmetrical cylindrical caldera [1].

4. Guamsan Tuff

Guamsan Tuff refers to a stratigraphic unit composed of volcanic breccias, ash-flow tuffs, fallout tuffs, and tuffites derived from the Guamsan caldera (**Figure 2**). The stratigraphic unit mostly consists of ash-flow tuffs that are only distributed inside the caldera (**Figure 1**). Though it was thick tuffs accumulated from the radial spreading of voluminous ash flow erupted from the crater hidden inside the caldera, it now remains only inside the caldera, due to prolonged erosion and denudation. The remaining body exposes its cross sections of the lower member (63.77 Ma in [9]) to upper member (60.1 Ma in [9]), because it has not only deep valleys that have been made by erosion but is also inclined northward by an igneous intrusion in the southern outer part of the caldera. The tuffs range 72~78% SiO_2 in composition, which indicates high silica to low silica [2].

The author describes lithofacies and mutual relations in the Guamsan Tuff and reconstructs the evolution processes of complex volcanic events from the volcanic ejecta.

4.1 Volcanic breccias

Volcanic breccias can be subdivided into two lithofacies of disorganized massive breccia and chaotic massive breccia.

The disorganized massive breccia mostly consists of monolithic blocks of rhyolite and accompanies rare accessory blocks of andesite and welded tuff.

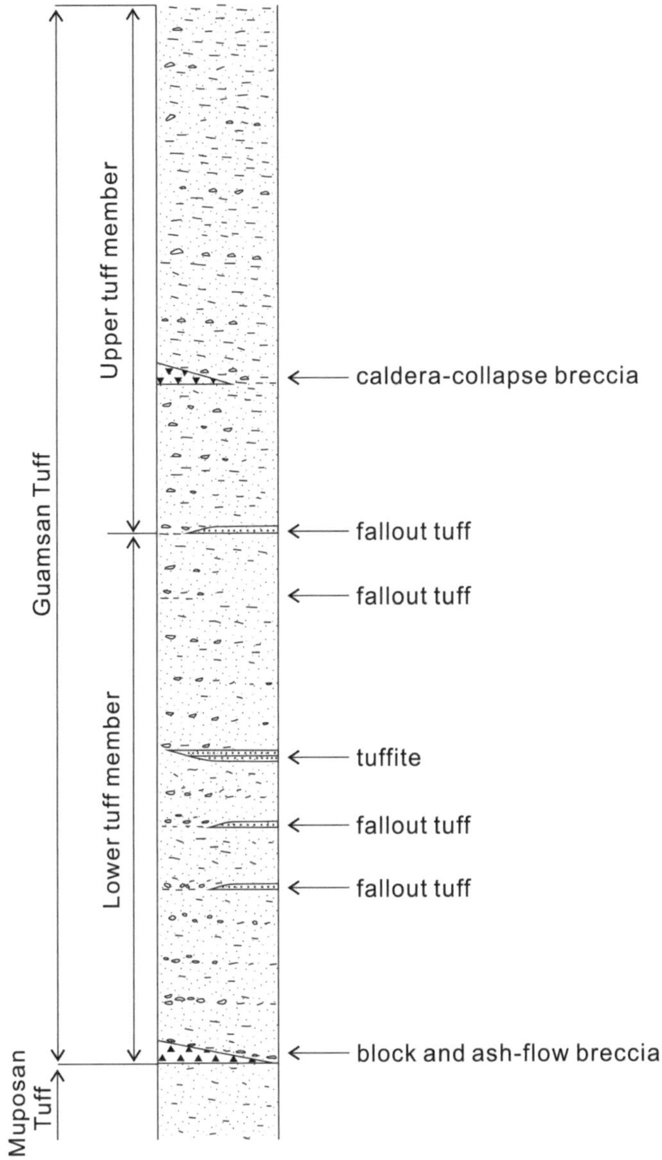

Figure 2.
Typical columnar section of the Guamsan tuff, which exhibits variations in stratigraphic units that comprise volcanic breccias, fallout tuffs, and tuffites in entire ash-flow tuffs.

The blocks are typically 5~15 cm in diameter, rarely reaching over 1 m, and are subround to subangular in shape (**Figure 3a**). The matrix consists of pale gray to gray ash that supports the blocks. The lithofacies exhibit the vertically thick lenticular form with laterally poor continuity. Typically, the lithofacies may be very thick in topographic depression and very thin on topographic high. Except for the blocks, it resembles lapilli tuff without internal stratification and grading. The lithofacies occur as basal breccia in the southwestern side of the outer ring dike (**Figure 4a**), from which the breccia gradually becomes grading into lapilli tuff in going northeastward. Along the emplaced site, it suggests that the breccia could be significantly different in terms of the mechanism forming the lithofacies. This can be supported by the fact that the trend is significantly different from those of other rock units in granulometric classification diagram.

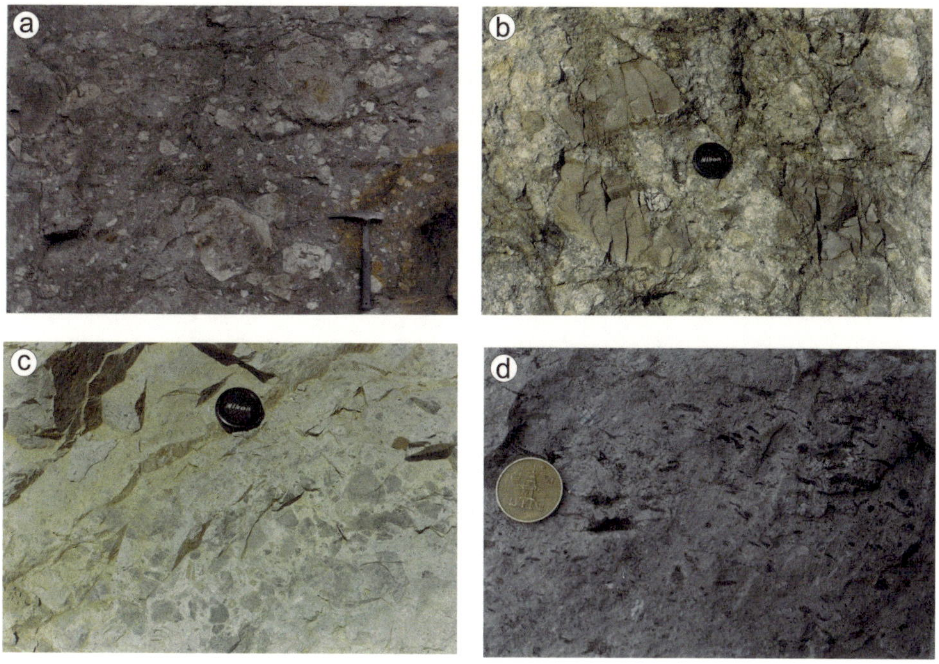

Figure 3.
Photographs in the Guamsan tuff. (a) Disorganized massive breccia in a near-vent facies; (b) chaotic massive breccia near the northern caldera margin; (c) graded lapilli tuff; (d) eutaxitic fabric in the middle part of the thick tuff and lapilli tuff bed.

The breccia embraces the southwestern side of the ring dike (**Figure 1**) and overlies the Naeyeonsan Tuff, cutting the Jugjang Volcanics. The lithofacies are correlated to the lowermost part of the Guamsan Tuff, based on sulfide alteration, stratigraphic relation, and lithologic correlation [1]. Therefore, the breccia may be considered as pyroclastic rocks accumulated from pyroclastic flows on a cone slope. The mechanism forming the pyroclastic flow may be dominated by the collapse of lava dome, because of the rhyolitic monolithic blocks. Namely, the pyroclastic flow-forming eruption type may be of the block and ash-flow phase that flowed along the slope from the collapse of active lava dome. Here, such lithofacies may indicate that it would be a near-vent facies, suggesting the initiation of volcanism in the Guamsan area.

Chaotic massive breccia, while it cannot be expressed on a geological map, is intercalated as a wedge shape between ash-flow tuffs in the lower part of the upper tuff member. The lithofacies consist of many blocks of rhyolite and welded tuff, which range 10~20 cm in diameter and occasionally over 1 m (**Figure 3b**). Based on granulometry, the rock is classified as tuff breccia, of which the trend is very different from those of other rock units. The boundary surface of the lithofacies can be investigated due to the fact that it is deeply eroded enough to expose the base. Though the base is irregular, the flat top is discontinuing laterally to be connected to normal lapilli tuffs. The lithofacies have very irregularly chaotic internal structures, which are cut by small faults (**Figure 3b**). The lithofacies, which occur sporadically in a valley adjacent to the northern ring fracture zone, are intercalated stratigraphically in the middle part of the Guamsan Tuff (**Figure 4b**). These facts suggest that the lithofacies had been produced by gravitational sliding of blocks or debris flow from caldera wall along the ring fracture zone. Therefore, many blocks around 1 m in diameter are observed around the collapse zone. Therefore, the breccia is considered as a caldera-collapse breccia, which is classified as the debris-flow phase by the caldera collapse. In addition, the lithofacies are thought to suggest a vent transition that indicates fissure eruption. If the eruption was initiated along the fissure, then it may have left many lithic fragments from the erosion of the vent walls. However, the content of lithic fragment decreases rapidly as it goes upward; this is a result of widening the conduit without erosion along caldera collapse due to the outward dipping of the fracture zone [1].

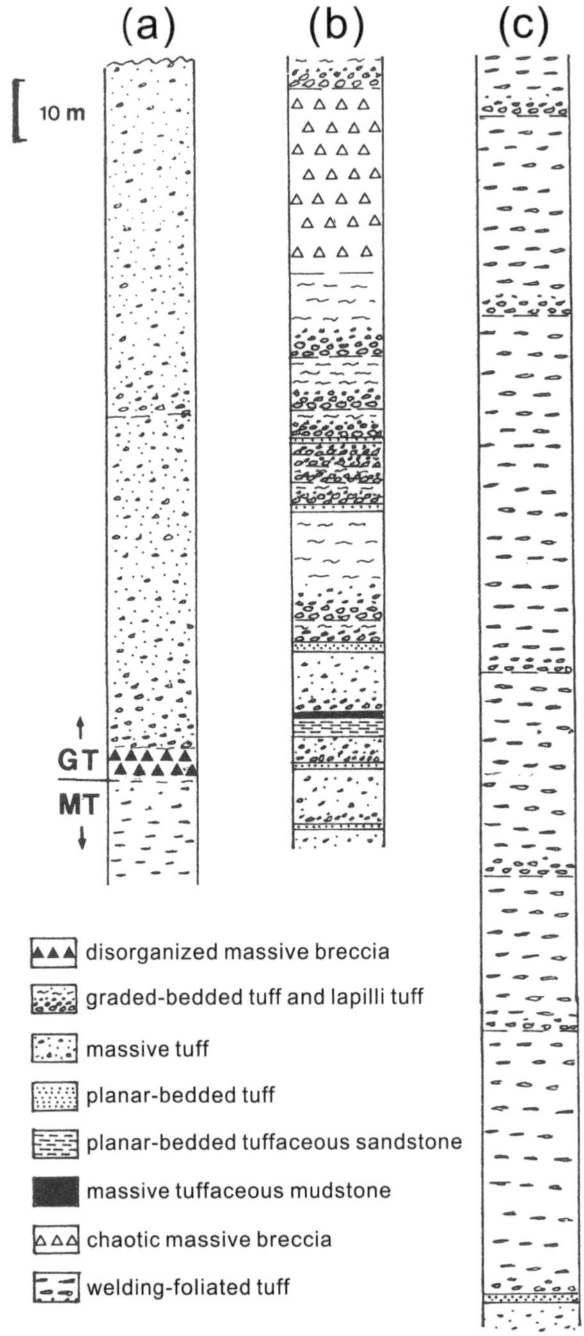

Figure 4.
Stratigraphic sections displaying variations in lithofacies of the Guamsan tuff (GT), underlain by the Muposan tuff (MT). (a) Lower member in the proximal zone; (b) lower member in the distal zone; (c) upper member of the Guamsan tuff.

4.2 Ash-flow tuffs

Lithofacies in the ash-flow tuffs comprises the graded tuff and lapilli tuff bed, massive tuff bed, and welding-foliated tuff bed.

The graded tuff and lapilli tuff beds exhibit whitish gray to pale gray color and are classified as lithic-rich vitric tuff consisting of lithics and pumice lapilli in the volcanic ash matrix. On the whole, the lithic lapilli are rich in the lower part, whereas pumice lapilli are more or less rich in the upper part. Therefore, lithics show a normal grading that gradually decrease upward in grain size within a single bed (**Figure 3c**). The matrix is composed of coarse ash, which is relatively richer than lapilli in the upper part. The lithofacies range from 2 to 20 cm in thickness and show laterally well extensibility. They represent eutaxitic fabric due to slightly welding in a thick single bed (**Figure 3d**), with lithics accumulated in its base (**Figure 5a**). The rock facies dominantly appear in the lower tuff member, whereas they occur in the lower part of the upper tuff member (**Figure 4a, b**). The grading in the lithofacies suggests a column collapse phase that was derived from the collapse of the high eruption column produced by a huge explosion. Then, collapsed tephra creates strongly fluidized pyroclastic flow that is supported by rising fluids in the same way as water vapor, etc., and dense lithics in flowing result in gradual sorting by the difference in final falling velocities.

Massive tuff bed is a lithic-rich vitric tuff that consists of lithics with various colors and a small amount of pumices. The matrix usually exhibits a whitish gray, pale gray, or pale bluish green color, whereas the lithics exhibit a dark gray, pale brown, or dark bluish green color; the pumices are mostly tinged with whitish gray color (**Figure 5b**). The matrix shows a massive appearance because it has very

Figure 5.
Photographs in the Guamsan tuff. (a) Lithic-rich zone in the lower part of the thick tuff and lapilli tuff bed; (b) pale gray massive tuff; (c) dark gray welding-foliated tuff in the upper member; (d) planar-bedded tuff beds intercalated between ash-flow tuff beds in the lower member.

abundant content and very poor sorting. The lithics exhibit very weak grading, with them accumulated at the base. The thickness ranges from 4 to 25 m in the bed, the boundary of which is distinguished by the accumulation of lithics in a single bed. The lithofacies mostly occur in the lower tuff member (**Figure 4a**) and are mostly plotted into the tuff field. Under a microscope, the tuff has a small amount of plagioclase and orthoclase phenocrysts, as well as extremely rare biotite and opaque minerals. Vitric shards appear intact as skeletal, crescent, and "Y" shapes in the lower part; their outlines sometimes appear by devitrification. Accordingly, they exhibit almost a vitroclastic fabric due to non-welding to partial welding in the lowermost part. However, the lower tuff member produces weak welding foliations by the gradual flattening of pumices and shards with an increase in welding degree, when going up the member. According to the poor sorting, massive bedding, very thick beds, and non-welding, it suggests that the lithofacies were emplaced as ash flows of tephra accumulated from the collapse of relatively high eruption column created by a little huge explosion. The weak grading of lithics signifies the weak fluidization of the ash flow. The abundance in matrix also suggests reduced escapement of volcanic ash from the ash flow due to weak fluidization. Therefore, the lithofacies indicate that they were emplaced by slightly fluidized ash flow from collapse of the high eruption column. However, the increase in welding degree going upward indicates a gradual decrease in the height of the eruption column as well as a gradual increase in the discharge volume.

The welding-foliated tuff bed occurs only in the upper tuff member (**Figure 4c**). The matrix, gray to dark gray in color, corresponds to a vitric tuff that includes a few of plagioclase without quartz grain. Although the tuff on the whole is not sorted and has no stratifications, it is welded so seriously that it shows welding foliation similar to lava together with any features representing high fluidity during its emplacement. Additionally, although the boundary of each flow unit could be determined because of the abundance in lithics at the base, the other boundaries may not be observed as they recede from almost complete welding. Commonly, despite the fact that the ash-flow tuff bed ideally has a surge tuff bed at the base (layer 1) and a fallout tuff bed on the top (layer 3), they are not discovered (**Figure 4c**).

The lithofacies include rare lithics, which are mostly lapilli 2~5 cm in size. Vitric ash and pumices are densely welded and have welding foliations (**Figure 5c**). Pumices exhibit a dark gray color and are extremely flattened by the thick superposition of many ash flows, but they are still difficult to detect due to their small size. They are not easily recognizable as they exhibit aspects similar to lava on fresh surfaces. However, because lithics in the lithofacies are, although small in size, concentrated on the base of flow units, they are used to distinguish the single flow units. Under a microscope, the lithofacies occur as a small amount of plagioclase and alkali feldspars as microphenocrysts and rarely quartz, biotite, and opaque minerals. The pumices are devitrified to be crystalized as micrographic fabrics of silica and feldspars or as mosaic fabrics by vapor-phase crystallization in the core. Therefore, although the lithofacies exhibit welded foliation by dense welding of pumices and shards, they show eutaxitic fabric in the lower part and parataxitic fabric in the middle part; they then transform into the vitrophyric fabric in the upper part, of which is difficult to recognize pyroclastic structure. That is, the tuff represents the vitrophyric fabric similar to obsidian showing no traces of devitrification in spite of the gradual increase in welding degree going upward.

The poor sorting and welding foliation indicate that the lithofacies were emplaced from ash flows. No occurrence of ash cloud-derived fallout tuff in any of the sections may reflect its location proximal to the eruption center or short emplacement time between flow units due to successive eruptions. The abundance in lithics in the base suggests that the lithics were lagged downward by the lateral movement of ash flows, even during a short time [10]. The sufficiently dense welding to have an indistinct boundary between fiammes and matrix suggests that they were significantly liquefied by almost complete welding under high temperatures. Based on the dense welding, as well as the sparsity or smaller size of phenocrysts in the upper part of the ash-flow tuff, the lithofacies indicate that they were emplaced by ash flows of relatively high temperature. Regarding the severe devitrification, the lithofacies indicate that the cooling period was relatively long by maintaining a longer time under high temperature by the reduction of heat loss, because of thick accumulation by rapid ash flows from the collapse of low eruption column.

The textural homogeneity, greater thickness, and denser welding reflect the sedimentary facies created by less fluidized non-expanded ash flows that have slow speed and less loss of volcanic ash into ash cloud, because they were originally erupted from hotter magma [11]. Such ash-flow phases also suggest that they originated from nonviolent voluminous eruptions of the type of boiling-over eruptions with continuous pulses. During the eruptions, it is thought that the flooding of repeated ash flows almost not occurring ash cloud had created so voluminous ash-flow tuffs that were accumulated inside the caldera. However, the earlier eruptions also produced ash-flow tuffs rich in lithics, because the ash flows occurred along ring fracture forming a caldera. Because the thick ash-flow tuffs helped longer preservation of high temperature in the caldera, they resulted in dense welding as well as high devitrification.

4.3 Fallout tuffs

Fallout tuff consists of pyroclastic rocks and tuffites. If the lithofacies are classified based on grain size and sedimentary structures, the pyroclastic rocks correspond to the planar-bedded tuff, whereas the tuffites comprise planar-bedded tuffaceous sandstone and massive tuffaceous mudstone.

Planar-bedded tuff is pale bluish green or gray in color and consists of medium to fine-grained ashes and extends laterally without any variations in thickness (**Figure 5d**). The lithofacies, which are about 1 m in thickness, are intercalated between ash-flow tuffs in medial or distal parts of northern and eastern margins of the caldera (**Figure 4b**). The lithofacies have poor sorting and normal grading that is finely grained upward in a single bed and rarely include accretionary lapilli ranging 5~10 mm in size, so they are easily recognized as a fallout tuff. No cross bedding or erosion tracks are discovered. The lithofacies is a vitric tuff that is mostly composed of vitric shards and scarce crystal grains. Crystal fragments consist of plagioclase, and quartzes are less than 1 mm in size and angular in shape. Evidences that the lithofacies are medium to fine in grain size and thinly intercalated between ash-flow tuffs indicate that the fallout tuff was derived from ash cloud following the ash flows. The fallout phase, derived from ash cloud, lies over the ash-flow tuff. Thus, the fallout tuff appears as a top facies following the normal facies of ash-flow tuff. Additionally, the lithofacies occurring in either medial or distal parts of northern and eastern margins of caldera suggests that their crater was located in the southwestern part of the caldera. Because the fallout phase derived from ash cloud becomes abundant as it goes away from the crater, it is almost not observable in the proximal part near the crater but most dominant in the distal part.

5. Postcollapse intrusions

Postcollapse intrusions in the Guamsan caldera are composed of rhyolitic dikes (60.65 Ma in [9]) and plugs, which are exposed as several lithofacies which are presumably connected to the identical magma chamber beneath the caldera. Based on differences in igneous structures and chemical composition, the lithofacies can be roughly subdivided into flow-banded rhyolite, porphyritic rhyolite, porphyritic rhyodacite, and stony rhyolite. In the field, a single rock body shows change in lithofacies within a few 100 m or a few km. In such cases, the erosion level in the Guamsan caldera area may nearly be approaching the original roof of the intrusions. Thus, according to positions related to the caldera-forming eruption, the intrusions can be categorized into intracaldera intrusions and ring dikes (**Figure 1**). The intrusions into the higher level of the Guamsan Tuff can demonstrate that they are the roots of postcollapse volcano, because the Guamsan Tuff has been of the final products and placed at a higher level as the caldera was formed [12].

5.1 Intracaldera intrusions

The intracaldera intrusions are intruding the Guamsan Tuff in the moat between the center and margin of caldera to form shapes of irregular circular plugs and straight dikes (**Figure 1**). In particular, they are annually distributed to form a circular ring shape along the caldera moat. Their exposure area is much wider in circular plugs than in straight dikes. The intrusions mostly consist of flow-banded rhyolite and rare spherulitic rhyolite in lithofacies. The rhyolite is mostly reddish gray in color and glassy and rarely contains plagioclase phenocrysts. Further, it develops flow foliations, especially spherulitic structure in the northern plug. The flow foliations have strikes almost parallel to intrusive contacts as well as steep dips ranging from 40 to 75°.

The lithofacies and occurrence patterns reflect that they are a vent region widened during the eruption of lava from residual magma rising through existing vents or fissures created due to crumpling by the collapse of the Guamsan caldera. In particular, the development of spherulitic structure in northern plugs is remarkable in contact with the fallout tuff. Therefore, the structure is to have been radially crystallized into vapor-phase crystallization by the moisture effect within the fallout tuff when the rhyolite intruded the tuff.

5.2 Ring dikes

Ring dikes occur along the caldera margin, and they can be categorized into the inner, intermediate, and outer ring dikes.

In terms of chemical composition, the dikes are mostly rhyolitic, whereas the southwestern ring dike is rhyodacitic in [2]. This indicates that the dikes are, chemically and mineralogically, similar to the Guamsan Tuff; this suggests that the ring dikes have a closely spatiotemporal relation with the caldera-forming eruption [13].

1. **Inner ring dike:** The dike is a combination of flow-banded rhyolite in the inner margin with stony rhyolite in the outer margin. The flow-banded rhyolite makes a steep slope in the inner margin of the ring dike. On the whole, the flow foliation is developed as closely spaced intervals and exhibits reddish gray to pale red colors. The flow-banded rhyolite is poor in phenocryst and alternates with very thin glassy bands and microcrystalline bands. The absence in phenocrysts seems to be the consequence of filter-pressing that passed only liquid except for crystals when magma was injected through fracture by compression in magma chamber. The glassy and microcrystalline textures indicate that the rhyolite was rapidly cooled because of the loss of volatile materials during magma rising instead of heat loss into conduit wall.

Stony rhyolite is frequently called felsite. It has a pale pink to pale gray color and indistinct or no flow foliation and rarely contains very tiny phenocrysts of quartz and feldspars. Under a microscope, quartz phenocrysts are rarely bipyramid in shape or embayed by resorption. The groundmass exhibits an intergranular texture by crystallization into microcrystalline to cryptocrystalline grain size. Therefore, according to the crystallinity, the earlier emplacement of inner side was followed by sequential intrusions after caldera collapse.

2. **Outer and intermediate dikes:** The outer ring dike intrudes along contact with the caldera and has contact with sedimentary rocks, Jugjang Volcanics, Naeyeonsan Tuff, and Muposan Tuff. The intermediate ring dike branches off inward from the southeastern part of the outer ring dike and then joins together at the south part; two small branches extend from the inner margin. The lithofacies exhibit whitish gray, pale gray, pale green, and pale pink colors, and they only show flow foliation in narrow dikes. They lack spherulite texture but represent stony or porphyritic textures. Thus, the outer ring dike shows stony rhyolite in the western and northern parts, porphyritic rhyolite in the eastern part, and porphyritic rhyodacite in the southern part, in which both are gradual. Further, the dike includes intrusive tuff in the northern part. In addition, the intermediate ring dike mostly consists of stony rhyolite.

Stony rhyolite contains microphenocrysts of plagioclase and alkali feldspars and accompanies tiny amounts of quartz and opaques. The phenocrysts are below 1 mm in diameter and are very small in content but also vary significantly depending on the locations. Occasionally, the rhyolite gradually converts into porphyritic rhyolite. The groundmass is mostly of quartzofeldspathic and occasionally contains biotite or opaques.

Porphyritic rhyolite contains phenocrysts of quartz, alkali feldspars, and plagioclase and rarely accompanies opaques. Quartz phenocrysts, up to 3 mm across, show embayed outline by significant resorption and contain rarely microcrystalline inclusions and occasionally bipyramid crystals. Some plagioclases constitute glomerophenocrysts. The groundmass shows an intergranular texture consisting of microcrystalline crystals. It is difficult to determine the boundary of such lithofacies because of the gradational change between porphyritic rhyodacite and stony rhyolite.

Porphyritic rhyodacite is dominant in plagioclase phenocryst and rarely accompanies alkali feldspars, hornblende, biotite, and opaques. Phenocrysts locally show very significant variations in content. The plagioclase phenocrysts are euhedral in shape and show a faint zonal structure. The groundmass consists of microcrystalline or cryptocrystalline feldspar, quartz, biotite, etc.

On the whole, the ring dikes show textural changes such as flow-banded, porphyritic, and stony structures in going toward the outer ring dike from the caldera center and also exhibit gradually compositional variation from rhyolite to rhyodacite in passing through southwest from the central part. Because all these constitute the ring dikes, this suggests that they intruded through ring faults or fractures resulting from caldera collapse.

It is thought that the emplacement timings may be locally different between rhyolites and various lithofacies. The textural changes from the caldera center to the outer ring dike suggest that the injection speed of magma was faster in the inner ring dike than in the outer ring dike, and the cooling rate was so slow that magma could crystallize to coarser grains owing to an intrusion of magma through wider passage in outer ring dike. Alternately, the changes between flow-banded textures and others may imply for temporal pulsational intrusions that the earlier injection of much hotter magma into the inner ring fault was followed by the injection of less hot magma into the outer ring fault, and changes between stony and porphyritic textures may also imply the sequential continuous intrusions of magma. However, the textural changes southwestward from the northern part of the caldera suggest that the erosion degree increased while going further southwestward.

The gradual compositional changes from the northern to southwestern parts of the caldera suggest in the sequential successive intrusion of magma that more silicic top of magma chamber first injected into the ring fracture zone, and then less silicic part of magma remaining down there injected into the zone sequentially. Therefore, the relationship between the two intrusions was the relationship between liquid and liquid. That is, the textural changes imply that rhyolite magma injected in the ring fracture was successively intruded by rhyodacite magma. The rhyolite magma was more evolved, with residual melt remaining sufficiently in the top of the magma chamber below the caldera block. First, the rhyolite magma was tapped in order to rapidly excavate fusible melt from the magma chamber, and then, second, the rhyodacite magma down there was successively tapped from the chamber [2].

6. Discussion

Interpretations of many outcrops in the Guamsan caldera area can reconstruct eruption types, caldera collapse, and volcanic processes. Now, the author will discuss the eruption types according to the volcanic processes involved with the caldera collapse. The Guamsan caldera area represents that the volcanic processes occurred in the following order: (1) Pyroclastic flow-forming eruption from column collapse, (2) caldera forming and vent shifting, (3) boiling-over eruption and ash-flow phase, (4) postcollapse intracaldera rhyolite intrusions, (5) postcollapse ring rhyolite dike intrusions, and (6) successive rhyodacite intrusion.

6.1 Pyroclastic flow-forming eruption from column collapse

The internal stratigraphic sequence of Guamsan Tuff can be analyzed so as to infer eruptions and their sequences. For cases of caldera-forming eruptions, the ideal internal sequence would comprise the following: (1) Plinian fallout eruption, (2) pyroclastic flow-forming eruption, and (3) effusive eruption. The initial Plinian eruption would turn into pyroclastic flow-forming eruption from collapsing the eruption column overloaded by the decrease in gas content, widening of the vent radius, etc. Therefore, the transition from fallout eruption to pyroclastic flow-forming eruption would be caused by the rapidly increasing discharge rate [14–16] and accompany the ejection of large volume of lithic fragments commonly forming coarse lag breccia [10, 17–19]. The transition connotes an initiation of new vents related to the caldera collapse [20]. However, not all pyroclastic flow-forming eruptions necessarily have an initial Plinian eruption [21, 22]. This may suggest that the vent radius was widened almost simultaneously with the collapse of vent area.

In the Guamsan area, it initiated with the Pelean eruption forming block and ash flows without the Plinian eruption. Evidences supporting an eruption include the block and ash-flow breccia, which possibly correspond to the lowest part of the Guamsan Tuff. The breccia that belongs to the pyroclastic breccia got accumulated on the slope of crater by pyroclastic flow. The mechanism forming the pyroclastic flow is dominated by the collapse of lava dome or eruption column. The fact that the lithofacies mainly consist of rhyolite blocks suggests that the Pelean eruption may have occurred by the temporary collapse of active lava dome, from which lava fragments flowed on a cone slope as block and ash flows. In addition, it turned into the eruptions forming strongly fluidized pyroclastic flow due to widening vent by collapse of the vent area. The eruption was probably a single vent phase erupted from a vent.

The lower member of Guamsan Tuff, overlying this breccia, consists of disorganized massive breccia, graded bedding tuff, lapilli tuff, block tuff, and sheet bedding tuff in lithofacies; the first three lithofacies correspond to pyroclastic flow rocks, while the last one belongs to fallout tuff. Because the fallout tuff mostly consists of ashes less than 1 mm in diameter, is less than 1 m thick, and is overlying each ash-flow tuff unit, it corresponds to the ash cloud-derived fallout phase rather than the Plinian fall phase. Thus, a pair of ash-flow tuff and fallout tuff represents the sedimentary phases of a flow unit and an ash cloud fallout unit that follow a pyroclastic flow. According to Sparks et al. [23], the flow unit corresponds to a main body of pyroclastic flow called layer 2, whereas the ash cloud fallout unit refers to layer 3. Here, the ground surge unit, called layer 1, is not recognized at the base. The ash-flow tuff without fallout tuff might be attributable to the successive pyroclastic flow that did not leave sufficient time to deposit fallout ash from ash cloud or to the rapid denudation of ash cloud-fallout deposits by fast violent pyroclastic flow near the crater. However, the pyroclastic flow-forming eruption successively occurred after the temporary Pelean eruption. Here, the mechanism forming pyroclastic flow was created by the collapse of the eruption column (**Figure 6a**).

The flow mechanism of pyroclastic flow represents partially fluidized flow [24–26]; the mechanism of transportation and sedimentation is similar to that of debris flow with slight differences. That is, while big clasts are transported by fine-grained matrix and water in the debris flow, big lithic fragments are transported by fine ashes and gas in the pyroclastic flow. Most pyroclastic flows are laminar flows in their body, although they are turbulent flows in their head. Pyroclastic flow can be distinguished into the three types of non-expanded flow, expanded flow, and segregating flow based on fluidization behaviors [11]. The ash-flow tuffs, occurring in the lower tuff member of Guamsan Tuff, seem to have been mostly emplaced by the expanded flow. The normal grading of lithic fragments and inverse grading of pumices in the lower ash-flow tuffs are dominated by settling velocity that depends on their density difference from matrix. The lithic fragments will occur normal grading according to each settling velocity because of the higher density of lithic fragments than matrix, whereas the pumices will occur the inverse grading according to floating over because of the lower density than matrix. Such grading appears more distinctly in accordance with the increase in density difference from matrix, when becoming greater in expansion degree of pyroclastic flow. The flow process of gas included during fluidization causes grading and sorting of fine-grained ashes. According to the expansion degree, the highly expanded flow would detach a large volume of fine-grained ashes from the flow to ash cloud, whereas the less expanded flow would drift upward only a small volume. Because the ash-flow tuffs in the lower tuff member mostly belong to the category of expanded flow, they represent not only the normal grading of lithic fragments but also planar-bedded tuffs by falling from the ash clouds detached from pyroclastic flows.

6.2 Forming caldera and vent transmitting

The ash-flow tuffs and fallout tuffs in the lower member were supplied from the central vent (**Figure 6a**). With the gradually increasing discharge volume of magma, the top of the magma chamber began sinking down like a ring shape. Here, the collapse occurred off northeastward from the central vent. The pressure in the magma chamber decreased gradually as the tapping of magma continued, and then the collapse seemed to occur, when the underpressure in the magma chamber exceeded the strength of overlying rocks. The roof collapse may recover lithostatic pressure at the top of the magma chamber. In this way, the role of central vent was rapidly reduced simultaneously with the initial collapse of caldera. The following

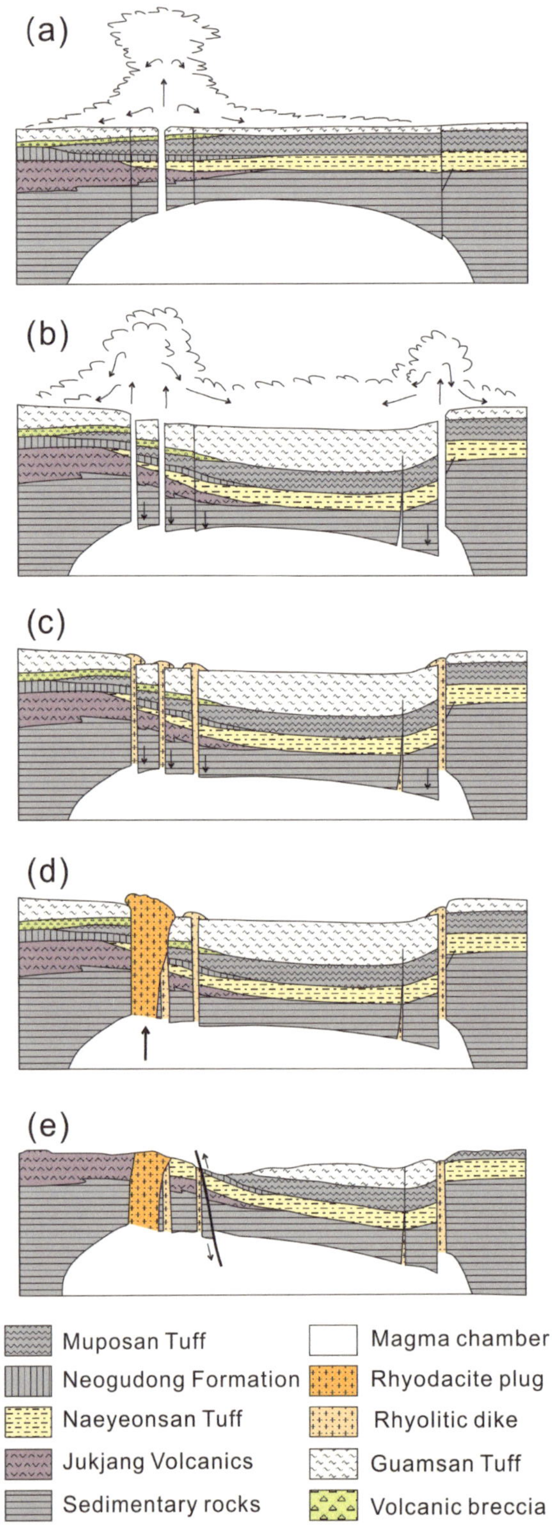

Figure 6.
Schematic sections, pictorially explaining the eruption types and volcanic processes in the Guamsan caldera. (a) Pyroclastic flow-forming eruptions through a central vent; (b) caldera collapse along ring fractures and boiling-over eruptions through multiple ring fissure vents; (c) effusive eruptions of rhyolitic magma through multiple vents along the intracaldera and ring fractures; (d) renewed rhyodacitic volcanism along the southwestern ring fracture; (e) present erosion surface.

ash flows were produced by erupting from many vents along the ring fracture zones, and then the caldera was collapsed further (**Figure 6b**). The pyroclastic flow-forming eruption with two-staged caldera-forming processes was already proposed by Druitt and Sparks [20].

The intercalation and lateral change of breccias in the Guamsan Tuff demonstrate that the collapse of Guamsan caldera is related with pyroclastic flow-forming eruption. This is directly evident in that the chaotic massive breccia is intruded as a wedge shape at the base of upper member of Guamsan Tuff. The breccia is a caldera-forming breccia produced by debris flow moving on steep caldera wall following the caldera collapse; this suggests the vent shift that connotes the fissure eruption. Thus, the appearance of breccia not only indicates the beginning with caldera collapse but also implies the closing of a single central vent as well as the initiation of multiple ring fissure vents. In addition, the Guamsan Tuff is nonexistent in extracaldera, whereas it is still over 850 m in thickness in intracaldera. Such thickness of the intracaldera tuffs is roughly three times thicker than the thickness of tuffs from extracaldera outflow, which is similar to cases of other tuffs from elsewhere around the world [12]. That is, this supports the fast accumulation of ash-flow tuff inside the caldera due to continuous eruption during the caldera collapse as opposed to collapsing the caldera after termination of the pyroclastic flow-forming eruption. Partial existence of intrusive tuff in northern ring dike can be counted as further evidence as well. In addition, although the lower member of Guamsan Tuff inside the caldera hardly exhibits welded zone but represents vitroclastic fabric, the upper member of Guamsan Tuff that exhibits welded foliation is dominant in eutaxitic and parataxitic fabrics and undergoing the devitrification process, leaving coarse-grained crystals. This is seemingly attributable to the high temperature that lasted for a long time by rapid accumulation of ash-flow tuff inside the caldera.

Thus, it is thought that the development of multiple ring fractures by caldera collapse during pyroclastic flow-forming eruption from central vent could serve as a natural momentum shifting to multiple vents along the ring fractures concurrently with the closing of a single vent.

6.3 Boiling-over eruption and ash-flow eruption

Some volcanologists have suggested from the two-staged models that the pyroclastic flow-forming eruptions converted into ring fracture vents from a single vent related with a caldera [17, 18], occasionally into the eruption from ring fractures [18, 27]. It seems to be the direct main cause that the shifting vents into ring fractures were transformed into the boiling-over eruption from the pyroclastic flow-forming eruption by column collapse. The conversion into boiling-over eruption was attributable to the rapidly increasing discharge of magma; the increase in discharge was possible from sinking the caldera block into magma chamber along the outward dipping of ring fracture zones. That is, because of the increasing number of vents together with outward dipping of ring fracture zones, the conduits were naturally gradually widened, and instantaneous discharge rate was rapidly increased as the subsidence occurred. Accordingly, the eruption at this moment did not form high eruption column but turned into the boiling-over eruption that produced a large volume of ash flow (**Figure 6b**). The discharge rate of magma at this moment was much higher than the discharge rate necessary for maintaining the eruption column.

In addition, the naturally widening conduits along subsidence were not necessary for eroding the conduit walls. Accordingly, the lithic fragments rapidly decrease in volume and are much smaller in size within the upper ash-flow tuffs as compared to the lower ash-flow tuff. It is very difficult to distinguish boundaries between each flow unit because of the fewer lithic fragments in the upper ash-flow tuffs, which are densely welded or even almost completely liquefied and devitrified enough to form eutaxitic and parataxitic fabrics. If it would form highly welded tuffs, it should be slow and less fluidized non-expansive ash flows, and it should be thickly accumulated during a short period, by the low column collapse derived from high-temperature magma, because heat loss could be prevented through the production and flooding of high-temperature ash flows. In order to reduce the height of the eruption column in this way, it is possible for magma to have a higher discharge rate, lower gas diffusion rate, and less gas content during eruption. Under such conditions, the ash flows may not spread far away and diminish with the loss of vitric ashes. Such pyroclastic flow-forming eruptions also indicate that they originated from nonviolent voluminous eruption, which supports that they were boiling-over eruptions that successively occurred without near explosiveness. The eruptions could produce voluminous ash-flow tuffs by flooding repetitive ash flows that do not almost produce ash cloud, whereas they did not have chances to accumulate fallout tuff.

The ring fissure eruption which caused such phenomena converted into effusive eruption together with the injections of magma into fissures by exhausting its explosive force and continuously collapsing caldera block due to the rapidly increasing eruption volume.

6.4 Intrusion of postcollapse intracaldera rhyolites

The boiling-over eruptions through ring fractures were also converted into non-explosive activities of magma by an exhaustion of volatile materials due to eruptions of voluminous ash flows. With the caldera formation, the residual magma were intruded into fissures of inverse wedge-like shape formed in the existing vent or in moat, and effusive eruptions successively occurred through the fissures (**Figure 6c**). The evidence for this is that the rhyolites occur in the moat as modes of a plug or dikes.

This also suggests that the distribution pattern of these rhyolites is dominant in fissures made in the moat along the caldera collapse. Thus, it is thought that the rhyolites are products of residual magma that filled fissures in the moat by the pressure imposed to the magma chamber after caldera collapse. Though volcanic domes formed at this time had already disappeared by erosion and denudation, it is presumable that they may have connected to the magma chamber as roots of postcollapse volcano with dome shape.

The following points of evidence suggest that they are intrusive roots of volcanoes after the collapse of caldera. Flow foliations in the rhyolite plugs dip almost vertically inside the caldera, whereas they dip toward the center outside the caldera. The intrusion of these rhyolites up to the high level of the Guamsan Tuff through fissures in the moat suggests that these are the roots of volcanoes after the collapse of caldera occurred at the higher level. The contacts of the rhyolites with Guamsan Tuff without cooling margin and the chemical composition of rhyolites similar to Guamsan Tuff also indicate that the intrusion of rhyolites was that the volcanic event occurred directly following the eruption of ash flows. That is, the intrusions can be regarded as underground residuals of volcanic activity occurring along the intracaldera fissures. At that time, the residual magma was probably very viscous and could not flow far away, fill only fissures up or at the upmost, and form small domes on the surface. Because they contain almost no phenocryst and develop flow foliation, they would perhaps have formed small lava domes on the ground.

6.5 Intrusion of postcollapse ring rhyolite dikes

Following the formation of the Guamsan caldera together with the ash flow-forming eruptions along the ring fracture, the residual magma formed ring rhyolite dikes by gradual injection into ring faults (**Figure 6c**). The injection of magma into ring fault, which is driven by the overpressure of magma causing its buoyance, can form dikes without any erosion of wall rocks [28]. The dikes are considered to be intrusion roots of several volcanoes created along the ring fracture after the caldera collapse. It is supposed that the volcanoes would have formed rhyolite domes aligned along the ring fracture on the ground. This is because the flow foliations in the ring dikes dip sub-vertically inside the caldera, whereas they dip inward outside the caldera.

At the same time as the intrusion of intracaldera rhyolites, the residual magma began with rising into ring fracture zones by the force accumulated therein. At that time, the residual magma was injected mainly into ring fissure vents, which were the passage of later ash-flow tuffs, and then formed the ring rhyolite dikes (**Figure 6c**). These are also a series of small postcollapse volcanoes that formed several rhyolite domes.

The rhyolites in ring dikes are generally glassy or microcrystalline in terms of crystallinity without cooling margins and distinct dragging in contact with adjacent rock body, implying that the dikes intruded the fissures along the ring fracture zones. That is, the ring dikes indicate that it is closely involved in the collapse of Guamsan caldera. Comparing the fact that the inner dikes are dominantly glassy with flow-banded to stony feature, the outer dikes are dominantly with stony to porphyritic texture and no cooling margins at the contact; this should be explained as a consequence of temporary pulsational intrusions, considering that magma chamber may be cooled down inward from the outside. However, sequential successive intrusion should be explained simultaneously, because the lithological relations in the outer and intermediate ring dikes are gradational. Such a relationship is strongly supported by the gradual compositional changes explained in the following.

6.6 Successive intrusion of rhyodacite

Following the intrusion of ring rhyolite dikes, the sequential successive intrusions of rhyodacite occurred in the junction site of outer and intermediate ring dikes in southwest part of the caldera (**Figure 6d**). The reason for this is that both dikes have the gradational lithology in the contact. Besides, the rhyodacites perhaps were intruded from the more crystallized inner part of the magma chamber which was cooled down inwardly from its margin and top. This is because, as compared to the rhyolite dike, which is mostly glassy to microcrystalline, the rhyodacite at the junction part exhibits a coarser texture that is porphyritic and microcrystalline.

The gradual change from rhyolite to rhyodacite in the outer ring dike suggests that the more silicic top of magma chamber was earlier injected along the ring fracture zone and following on it the more mafic magma below it was successively intruded. The emplacement of rhyodacite in the ring dike was possible by discharging the rhyodacite magma following after exhaustion of effective melt by tapping the rhyolite magma [2]. By accounting for the patterns and locations of the ring dikes, the dikes can be judged to be the products of magma rising along the junction part of two ring fracture zones during the final stage of Guamsan volcanic activities; thus, they are also regarded as the roots of postcollapse volcanoes connected to the magma chamber.

The intrusion of this rhyodacite resulted in slow crystallization after rising magma emplaced through the junction part of ring dikes; this implies the closure of the Guamsan volcanic activities. Therefore, this spatiotemporal view corresponds to a final intrusion along the ring fracture zone of southwestern caldera in Guamsan magmatic system, but they now display their root zones due to deep erosions during long time (**Figure 6e**).

7. Conclusions

Stratigraphic units associated with the Guamsan caldera comprise Guamsan Tuff and rhyolitic intrusions. Guamsan Tuff typically consists of volcanic breccia, ash-flow tuffs, and fallout tuffs, of which the ash-flow tuff is very dominant.

Volcanic breccia can be distinguished into the block and ash-flow breccia at the lower part and the caldera-collapse breccia at the upper part, according to their distribution locations and stratigraphic sequence. In the lower member, the ash-flow tuffs exhibit expansive pyroclastic flow phase by the pyroclastic flow-forming eruptions, and the fallout tuffs exhibit ash cloud-falling phase, whereas the ash-flow tuffs in the upper member exhibit the non-expansive ash-flow phase by the boiling-over eruptions.

Rhyolite intrusions can be distinguished into intracaldera intrusions and ring dike based on location and pattern: the ring dikes are distinguished into the inner, intermediate, and outer ring dikes. The Guamsan caldera represents a caldera cycle connecting into ash-flow tuff—caldera—ring dikes.

The volcanic processes in the Guamsan caldera area can be summarized as the following sequence along the caldera cycle: (1) The volcanic activity began with the Pelean eruption generating block ash-flow phase by a lava dome, and (2) it then subsequently turned into the fluidized pyroclastic flow phase by the collapse of high eruption column. At that time, fluidization of the pyroclastic flow was reduced with gradual decrease in the height of the eruption column. (3) In the transformation into ash-flow phase, boiling-over eruptions burst instantaneously hotter pyroclastic materials to be densely welded. The boiling-over eruptions began on their way by migrating vents into ring fracture zones together with the caldera collapse. At the earlier stage of eruption, the pyroclastic flows were produced from a central vent, whereas voluminous ash flows were generated from multiple vents in the ring fracture zone. The consequently accumulated Guamsan Tuff was at least 850 m thick inside the caldera. (4) After the ash-flow eruptions, the magma was injected into fissures in the caldera moat to form the rhyolitic plug and dikes. (5) Almost simultaneously, the magma was successively injected along the ring fracture zone so as to form the ring rhyolite dikes. (6) Finally, the rhyodacite was successively intruded into the junction part of southwestern ring dikes.

Acknowledgements

The work was supported by funding of the Korea Meteorological Administration Research and Development Program under Grant KMI (**2018-01610**) through the Korea Meteorological Institute. We are grateful for the careful approval for our proposal by the editor and reviewer, Professor Angelo Paone.

Author details

Sang Koo Hwang
Department of Earth and Environmental Sciences, Andong National University, Andong, Korea

*Address all correspondence to: hwangsk@anu.ac.kr

References

[1] Hwang SK. Collapse type and evolution of the Guamsan caldera, southeastern Cheongsong, Korea. Journal of the Geological Society of Korea. 2002;**38**:199-216

[2] Hwang SK. Magmatic evolution of volcanic rocks related with the Guamsan caldera, southeastern Korea. Journal of the Geological Society of Korea. 2002b;**38**:293-310

[3] Lee SM, Kim SW, Jin MS. Cretaceous to tertiary volcanic activities and tectonic significance of South Korea. Journal of the Geological Society of Korea. 1987;**23**:338-359

[4] Hwang SK, Kim SW. Petrology of cretaceous volcanic rocks in the Milyang-Yangsan area: Petrotectonic setting. Journal of the Geological Society of Korea. 1994;**30**:229-241

[5] Chough SK, Kwon S-T, Ree J-H, Choi DK. Tectonic and sedimentary evolution of the Korean peninsula: A review and new view. Earth-Science Reviews. 2000;**52**:175-235

[6] Hwang SK. Tectonic setting and arc volcanisms of the Gyeongsang arc in the southeastern Korean peninsula. The Journal of the Petrological Society of Korea. 2012;**21**:367-383

[7] Hwang SK. SHRIMP U-Pb dating and volcanic history of the Jipum Volcanics, western Yeongdeok, Korea. The Journal of the Petrological Society of Korea. 2017;**26**:341-352

[8] Hwang SK, Jo IH, Yi K. SHRIMP U-Pb zircon dating and stratigraphic relationship of the Bunam stock and Muposan tuff, Cheongsong. Journal of the Geological Society of Korea. 2016;**52**:405-419

[9] Hwang SK, Jo IH, Yi K. SHRIMP U-Pb dating and volcanic processes of the volcanic rocks in the Guamsan caldera, Korea. Economic and Environmental Geology. 2017;**50**:467-476

[10] Wright JA, Walker GPL. The ignimbrite source problem: Significance of a co-ignimbrite lag-fall deposits. Geology. 1977;**5**:729-732

[11] Wilson CJN. The role of fluidisation in the emplacement of pyroclastic flows: An experimental approach. Journal of Volcanology and Geothermal Research. 1980;**8**:231-249

[12] Lipman PW. Roots of ash-flow calderas in western North America: Windows into the tops of granitic batholiths. Journal of Geophysical Research. 1984;**89**:8801-8841

[13] Smith RL, Bailey RA. Resurgent cauldrons. Geological Society of America, Memoir. 1968;**116**:613-662

[14] Bursik MI, Woods AW. The dynamics and thermodynamics of large ash flows. Bulletin of Volcanology. 1996;**58**:175-193

[15] Hwang SK, Lee GD, Kim SW, Lee YJ. Volcanisms and volcanic processes of the Wondong caldera, Korea. The Journal of the Petrological Society of Korea. 1997;**6**:96-110

[16] Hwang SK, Kim SW, Lee YJ. Volcanisms and igneous processes of the Samrangjin caldera, Korea. The Journal of the Petrological Society of Korea. 1997;**7**:147-140

[17] Druitt TH. Vent evolution and lag breccia formation during the Cape Riva eruption of Santorini, Greece. Journal of Geology. 1985;**93**:439-454

[18] Druitt TH, Bacon CR. Lithic breccia and ignimbrite erupted during the collapse of Crater Lake caldera, Oregon.

Journal of Volcanology and Geothermal Research. 1986;**29**:1-32

[19] Rosi M, Vezzoli L, Aleotti P, De Censi M. Interaction between caldera collapse and eruptive dynamics during Campanian ignimbrite eruption, Phlegrean fields, Italy. Bulletin of Volcanology. 1996;**57**:541-554

[20] Druitt TH, Sparks RSJ. On the formation of calderas during ignimbrite eruptions. Nature. 1985;**310**:679-681

[21] Francis PW, O'Callaghan L, Kretzchmar GA, Thorpe RS, Sparks RSJ, Page RN, et al. The Cerro Galan ignimbrite. Nature. 1983;**301**:51-53

[22] Hwang SK, Kim SW. Silicic volcanism of Yangsan caldera, Korea. Journal of the Geological Society of Korea. 1992;**28**:491-503

[23] Sparks RSJ, Self S, Walker GPL. Products of ignimbrite eruptions. Geology. 1973;**1**:115-118

[24] Sparks RSJ. Grain size variations in ignimbrites and implications for the transport of pyroclastic flows. Sedimentology. 1976;**23**:147-188

[25] Sparks RSJ, Bursik MI, Carey SN, Gilbert JS, Glaze LS, Sigurdsson H, et al. Volcanic Plumes. Chichester: Wiley; 1997. p. 574

[26] Branney MJ, Kokelaar P. Pyroclastic density currents and sedimentation of ignimbrites. Geological Society, London, Memoirs. 2002;**27**:143

[27] Hildreth W, Mahood GA. Ring-fracture eruption of the bishop tuff. Geological Society of America Bulletin. 1986;**97**:396-403

[28] Gudmundsson A, Marinoni LB, Marti J. Injection and arrest of dykes: Implications for volcanic hazards. Journal of Volcanology and Geothermal Research. 1999;**88**:1-13

Chapter 5

Toward a New Conceptual and Methodological Approach for the Integral Evaluation of Volcanic Risk

Leonel Vega

Abstract

In the world, there are very few experiences of studies oriented to the integral evaluation of risks due to natural hazards. In the case of volcanic risk, most of the scientific-technical and economic efforts have been oriented mainly toward the evaluation of threats, with few methodological considerations to assess vulnerability and much less risk. In other cases, the threat and vulnerability are evaluated independently, with many difficulties for the comprehensive risk assessment. Many of the studies called "vulnerability assessments" are only physical and functional characterizations and diagnoses of vital infrastructure and population. These characterizations can hardly be interpreted in terms of georeferenced indices and/or vulnerability maps that represent the spatial and temporal exposure of the elements exposed to each threat and, even less, that represent the intrinsic and extrinsic response capacities of these elements in comparison with the threats. In this chapter, a new conceptual and methodological approach is proposed for the integral evaluation of volcanic risk, which includes the generation and adjustment of a new equation for the determination of volcanic risk, based on the integral assessment of threats and vulnerabilities.

Keywords: volcanic threat, vulnerability, volcanic risk index, intrinsic and extrinsic response capabilities, maps of risk

1. Introduction

So far, there is no technique to accurately predict the occurrence of a volcanic eruption. Some of the phenomena presented by volcanoes such as seismic activity (tremors, etc.), soil deformation, gas emissions, or fumarolic activity and the chemical composition of water and its vapors help scientists to know when a volcano begins to activate. If changes in these phenomena can be detected, it is possible to establish some degree of probability of a volcanic eruption, although it is impossible to predict the day, time, and size of an eruption [1].

To detect these changes, volcanic observatories have been installed in various volcanoes around the world for several years, equipped with a series of equipment that has been collecting valuable information, allowing in some cases to predict at least the time when the activity would begin in the surface and the place where the materials would be emitted [2–4]. Anyway, volcanoes have individual behaviors, so it is necessary to make a permanent and specific follow-up to each one, because although it is true there are some features common to all, there are others that individualize them [5–7].

It is also important to investigate the history of each volcano through the identification, petrographic analysis, and dating of its multiple pyroclastic deposits, to determine the characteristics that typify them [8, 9]. With this information and other knowledge, it is possible to elaborate, for example, maps of volcanic threats, which although they do not allow to determine when the next eruption will be, if they allow to determine an approximate order of the magnitude of the event and of the areas of affectation [10].

In fact, in the world, there are very few experiences of studies oriented to the integral evaluation of risk in the face of natural hazards. So much so that in the case of volcanic risk, most of the scientific-technical and economic efforts have been oriented mainly toward the evaluation of threats, with few methodological considerations for the evaluation of vulnerability and much less of the risk [11]. In other cases, the threat and vulnerability are evaluated independently, which logically presents many difficulties for the integral risk assessment. It is also easy to verify that many of the studies called "vulnerability assessments" are only physical and functional characterizations and diagnoses of vital infrastructure and population [12–14]. These characterizations can hardly be interpreted in terms of georeferenced indexes and/or maps of vulnerability that represent the spatial and temporal exposure of the elements exposed to each threat, much less that they represent the intrinsic and extrinsic response capacities of these elements compared to the threats.

What is required, then, is to define to whom and to what this event could affect, its degree of vulnerability to the threat, and the level of risk to which it is subjected, as basic inputs for decision-making and comprehensive risk management.

In this chapter of book, in light of the process of "Systemic Parametrization of the Environmental Dimension" [15], a summary of the conceptual and methodological approach developed by the undersigned is presented through the PIGA Group for Research in Politics, Information, and Management Environmental of the Universidad Nacional de Colombia, to carry out the studies and analysis of vulnerability and risk in a sector of the area of influence of the Cerro Machín volcano [16], taking as a starting point the study of the volcanic threat previously advanced by the former Colombian Institute of Geology and Mining [17].

Finally, some general conclusions and recommendations are presented with the hope that this new approach constitutes another grain of sand in the difficult task of protecting human beings and their environment from natural threats, particularly from volcanic threats, all through of an integral management of the risk that evaluates and anticipates the threats in a timely manner, that adequately plans and budgets the policies, strategies, instruments, and protocols to be followed in front of them, and that responds with effectiveness against the handling of emergencies and contingencies. In any case, it is expected to understand that "there are no natural disasters but political and management disasters."

2. New conceptual approach

Traditionally, the definition of risk (R) refers to the probability that something harmful will happen on a given element [18]. The simplest conceptual expression to express the risk has been R = A. V, where A is the threat, understood as a latent condition derived from the probability of occurrence of a physical phenomenon of natural, socio-natural, or anthropic unintentional origin that it can cause damage to the element or group of exposed elements, and V is the vulnerability, understood as the susceptibility or characteristic of the element or group of elements to be totally or partially damaged by the impact of the threat [19].

In the development of this chapter of book, a new conceptual and methodological approach is proposed for the integral evaluation of volcanic risk, which includes the generation and adjustment of a new equation for the determination of volcanic risk, based on the integral assessment of threats and vulnerabilities.

2.1 Threat analysis

Consistent with [18, 19], the threat represents the potential for damage of a natural phenomenon and is calculated by quantifying the energy that is applied to a particular site of interest or unit of analysis.

For the purposes of this study, it is assumed that the energy of a threat (as well as that of an environmental impact) can be represented qualitatively according to its intrinsic characteristics of probability of occurrence, intensity, duration, extension, accumulation, synergy, etc. [20], and, therefore, the quantification of this energy is done by means of an index that represents dimensionally and under the same scale the intrinsic characteristics of the different volcanic threats considered.

Consequently, taking as reference, the equation model that calculates the intrinsic importance in environmental impacts [20], the intrinsic threat index (Å) is calculated for each threat j of each analysis scenario based on its main intrinsic characteristics as shows in Eq. (1):

$$\mathring{A}_j = P\,(\,0{,}6.I_j + 0{,}2.D_j + 0{,}1.E_j + 0{,}1.A_j\,) \tag{1}$$

where Å is the intrinsic threat index, P is the probability of occurrence, I is the intensity of the threat, D is the duration of the threat, E is the extension of the threat, and A is the accumulation of the threat.

For the qualitative assessment of each of the characteristics that determine the intrinsic threat index, the environmental impact assessment model is taken as a reference [20], and **Table 1** is generated where the different assessment categories are proposed.

INTRINSIC THREAT	Threat index calculated for each threat j in the units of analysis. It depends on its probability of occurrence and its intrinsic characteristics of intensity, duration, extension and accumulation.	Irrelevant	0 - 0.24
		Moderate	0.25 - 0.49
		Severe	0.50 - 0.74
		Critical	0.75 - 1.00
PROBABILITY OF OCCURRENCE	Probability of occurrence of the threat	Nula	0
		Total	1
INTENSITY	The amount of energy that manifests itself, in the unit of analysis, is related to the destructive potential of the threat	Low	0,001
		Media	0,01
		High	0,1
		Very High	1
DURATION	Duration of the threat occurring, in the unit of analysis. Immediate (less than hours), medium term (hours to days), long term (days to weeks)	Righ now	0,01
		Medium term	0,1
		Long term	1
EXTENSION	Coverage that has the threat in the unit of analysis	Nula	0
		Partial	0.1-0.9
		Total	1
ACCUMULATION	A threat is cumulative if it manifests itself in the unit of analysis several times and this increases its potential for damage.	Simple	0,1
		Cumulative	1

Table 1.
Valuation of the intrinsic threat index.

2.2 Vulnerability analysis

For the purposes of this study, vulnerability will be associated with the ability of an element or group of elements not to be totally or partially damaged by the impact of a threat [21]. Conceptually, it will be a function of the degree of spatial and temporal exposure, and of the intrinsic and extrinsic response capacity of the exposed elements.

In order to be able to mathematically integrate the intrinsic threat index (Å) with the vulnerability values, with the help of the Excel tool and after successive tests with field information, Eq. (2) is generated and adjusted for the vulnerability index (V), which is calculated for each exposed element i against each threat j, as described below:

$$V = SE.TE.(1 - IRC)^{1+\alpha.ERC} \qquad (2)$$

where V is the vulnerability index, SE is the space exhibition, TE is the temporary exhibition, IRC is the intrinsic response capacity, ERC is the extrinsic response capacity, and α is the form coefficient used in the adjustment of the family of curves corresponding to the vulnerability Eq. (2) (see **Figure 1**).

For the qualitative assessment of each of the characteristics that determine the vulnerability index, **Table 2** is generated, where the different assessment categories are proposed.

2.2.1 The intrinsic response capacity (IRC)

For the purposes of this study, the intrinsic response capacity (IRC) will be understood as an index that represents dimensionally the capacity of each exposed element (ecosystem, constructed, population) to react and/or physically resist the impact of a threat and/or recover later by itself from the affectation caused.

The IRC is based on the concept of resilience, whose definition of the term comes from the field of physics, referring to "the ability of a material to recover its original form after having been subjected to high pressures," and that in its broadest sense, it is described as "elasticity" [11]. Later, due to multiple similarities and analogies, the concept of resilience extended to the field of natural and social systems but, in any case, always denoting "the degree to which a system recovers or returns to its previous state before the action of an external stimulus" [12].

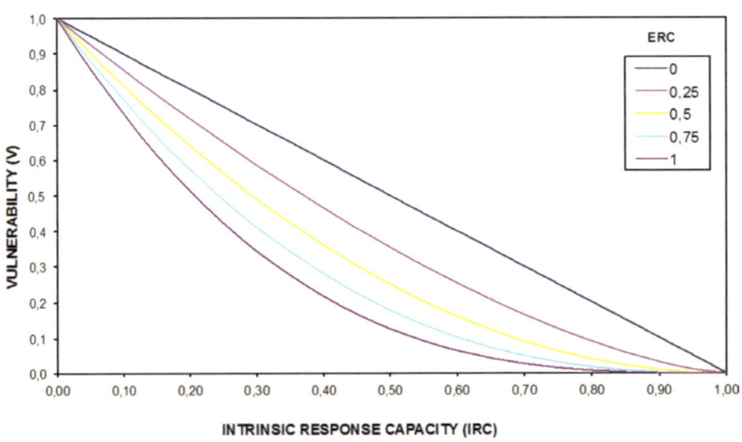

Figure 1.
Family of curves in the vulnerability equation (V).

V	VULNERABILITY	Vulnerability Index, calculated for each exposed element i of the environment against each threat j. It is a function of the degree of spatial and temporal exposure and the intrinsic and extrinsic response capacity of the exposed elements.	Irrelevant	0 - 0.24
			Moderate	0.25 - 0.49
			Severe	0.50 - 0.74
			Critical	0.75 - 1.00
SE	SPACE EXPOSURE	Degree of spatial exposure of the element to the threat. It is classified according to the space or area exposed to the threat classified as a percentage, for which the GIS is used.	No one	0,00
			Low	0,25
			Media	0,50
			High	0,75
			Very high	1,00
TE	TEMPORAL EXPOSITION	Generally the temporary exposure corresponds to the useful life of the element. For deterministic scenarios, it corresponds with the time the exposed element lasts when the threat occurs.	No one	0,00
			Hours to days	0,25
			Days to weeks	0,50
			Weeks to months	0,75
			Permanent	1,00
IRC	INTRINSIC RESPONSE CAPACITY	The capacity of the exposed element (ecosystem, built, population) to react and physically resist the impact of a threat and to subsequently recover by itself from the affectation caused.	No one	0,00
			Low	0,25
			Media	0,50
			High	0,75
			Very high	1,00
ERC	EXTRINSIC RESPONSE CAPACITY	Institutional capacity to comprehensively manage risk in compliance with the basic systemic functions of planning, management and evaluation.	No one	0,00
			Low	0,25
			Media	0,50
			High	0,75
			Very high	1,00

Table 2.
Assessment of the vulnerability index.

Consequently, the IRC will depend on each type of threat in particular and will be calculated independently for each element exposed based on a weighted assessment of attributes, according to the generic Eq. (3):

$$IRC = \frac{\Sigma Pn.Wn}{Pn_{m\acute{a}x}} \tag{3}$$

where IRC is the intrinsic response capacity, Pn is the evaluation of attributes according to characteristics of each exposed element, and Wn is the weighting factor.

The intrinsic ecosystem response capacity (ICRe) is defined as the capacity of an ecosystem to react and physically resist the impact of a threat and subsequently recover by itself from the damage caused. It depends on each type of threat in particular and can be calculated independently for each exposed element of the ecosystem (rivers, páramos, forests and stubble, pastures, and crops) based on a weighted assessment of descriptors and attributes related to the environmental state of the ecosystems, in terms of quantity, quality, and ecological availability of environmental goods and services, and the degree of intervention or anthropic pressure, in terms of the use and deterioration caused on said environmental goods and services [16, 20].

The intrinsic response capacity of constructed elements (IRCc) is defined as the capacity of a constructed element to physically resist the impact of a threat and to maintain its functionality after the affectation received. It depends on each type of threat in particular and can be calculated for each exposed constructed element (buildings, roads, infrastructures) based on the weighted assessment of descriptors and attributes related to their physicochemical characteristics such as construction material (from the structure, elements, base, subbase), the structure (type, mezzanines, anchors), the roof (type of roof), the covering (type of covering), the rolling (rolling layer), the terrain (ground, slope), drains (quantity and condition of drainage works), and general condition (age, conservation, damage) [5, 14, 16].

The intrinsic response capacity of the population (IRCp) is defined as the capacity of a given population to react and physically resist the impact of a threat and subsequently recover by itself from the affectation caused. It can be calculated for an exposed population group based on a weighted assessment of descriptors and attributes related to planning (perception of risk, level of education, unsatisfied basic needs, participation in drills, participation in emergency committees, knowledge of evacuation routes and shelters), the operation (optimal evacuation distance, type and quality of route, population to be mobilized, active and passive human resources, physical and/or psychological limitations), and logistics (means of transport and communication equipment) [11, 12, 16].

2.2.2 The extrinsic response capacity (ERC)

For the purposes of this study, the extrinsic response capacity (ERC) will be understood as an index that represents dimensionally the institutional capacity of the entities responsible for the integral management of the risk of responding orderly and efficiently to emergency situations that generate one or more threats determined [16]. It does not depend on the threats, and therefore it is calculated for each exposed population group (country, department, municipality, township, village) according to the generic Eq. (4):

$$ERC = \frac{\Sigma Pn.Wn}{Pn_{máx}} \tag{4}$$

where ERC is the extrinsic response capacity, Pn is the assessment of attributes of institutional capacity, and Wn is the weighting factor.

In accordance with the general functions of an incident command system (ICS) [22], the following descriptors and attributes for the ERC are proposed:

Planning: identification and characterization of risks, emergency plans, availability evacuation routes and shelters, simulation programming and coordination, and conformation and coordination of emergency committees.

Operation: optimal assistance distance, type and quality of route, population to be assisted, social care, medical assistance, and technical assistance in search and rescue.

Logistics: availability and management of supplies, communication system and early warning, transport, and facilities and equipment.

2.3 A new risk equation

As suggested, for the purposes of this study, comprehensive risk assessment is a process with a holistic, systemic, and environmental approach [16, 20], and, therefore, the definition of risk (R) refers to the probability that something harmful can happen in a certain environment or in a segment or element of it (ecosystem, public sector, economic sector, civil society).

In this context, with the help of the Excel tool and after successive trials with field information and conceptual and methodological approaches that avoided the null values for threats and vulnerabilities, a new expression was adjusted, as an index, for the determination of risk against volcanic threats, as shown in Eq. (5):

$$R = \text{Å}^{a}.V^{b} \tag{5}$$

where R is the risk index, Å is the intrinsic threat index, V is the vulnerability index, a is [b - c. ln (V)], b and c are the shape coefficients in the fit of the family of curves corresponding to the risk Eq. (5), as shown in **Figure 2**.

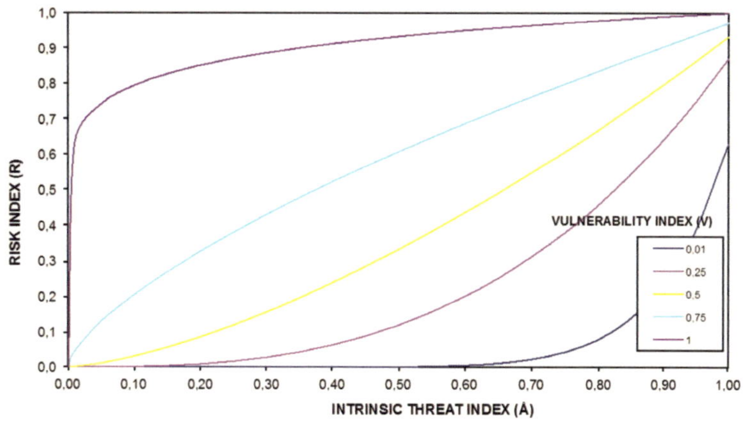

Figure 2.
Family of curves in the risk equation (R).

3. Methodological approach

The proposed methodological approach for comprehensive risk assessment involves two fundamental elements, the logical framework matrix and the process diagram, as explained below.

3.1 The logical framework matrix

Taking as a reference the logical framework matrix for systemic and integral evaluation of environmental impacts proposed in [20], the logical framework matrices are designed and defined for the integral evaluation of volcanic risk in the scenarios of the onset of crisis and eruption, which is shown in **Tables 3** and **4**.

3.2 The process diagram

Figure 3 schematizes the process diagram proposed for the integral evaluation of volcanic risk, which is consistent with the previously described conceptual framework.

To apply and develop this methodology is essential to have GIS tools [23], whose specific process includes a series of activities such as the collection and structuring

Exposed Elements	P [0-1]	CHARACTERIZATION OF THREATS AND VULNERABILITIES				Total risk on each exposed element
		Earthquakes		Landslides		
Ecosystem	P_e	$Å_s$ / V_{es}	$R_{es} = f(Å_s, V_{es})$	$Å_d$ / V_{ed}	$R_{ed} = f(Å_d, V_{ed})$	$R_e = \sum R_{ej} / \sum R_{(ej)máx}$
Constructed	P_c	$Å_s$ / V_{cs}	$R_{cs} = f(Å_s, V_{cs})$	$Å_d$ / V_{cd}	$R_{cd} = f(Å_d, V_{cd})$	$R_c = \sum R_{cj} / \sum R_{(cj)máx}$
Population	P_p	$Å_s$ / V_p	$R_{ps} = f(Å_s, V_p)$	$Å_d$ / V_p	$R_{pd} = f(Å_d, V_p)$	$R_p = \sum R_{pj} / \sum R_{(pj)máx}$
Total risk for each threat		$R_s = \sum (R_{is} . P_i)$		$R_d = \sum (R_{id} . P_i)$		$Rt = \sum (Ri . Pi)$

Table 3.
Logical framework matrix for the integral risk assessment—start crisis scenario.

Exposed Elements	P [0-1]	CHARACTERIZATION OF THREATS AND VULNERABILITIES				Total risk on each exposed element
		Falls	Flows	Earthquakes	Landslides	
Ecosystem	P_e	$\frac{A_k}{V_{ek}}$ $R_{ek} = f(A_k, V_{ek})$	$\frac{A_f}{V_{ef}}$ $R_{ef} = f(A_f, V_{ef})$	$\frac{A_s}{V_{es}}$ $R_{es} = f(A_s, V_{es})$	$\frac{A_d}{V_{ed}}$ $R_{ed} = f(A_d, V_{ed})$	$R_e = \sum R_{ej} / \sum R_{(ej)max}$
Constructed	P_c	$\frac{A_k}{V_{ck}}$ $R_{ck} = f(A_k, V_{ck})$	$\frac{A_f}{V_{cf}}$ $R_{cf} = f(A_f, V_{cf})$	$\frac{A_s}{V_{cs}}$ $R_{cs} = f(A_s, V_{cs})$	$\frac{A_d}{V_{cd}}$ $R_{cd} = f(A_d, V_{cd})$	$R_c = \sum R_{cj} / \sum R_{(cj)max}$
Population	P_p	$\frac{A_k}{V_p}$ $R_{pk} = f(A_k, V_p)$	$\frac{A_f}{V_p}$ $R_{pf} = f(A_f, V_p)$	$\frac{A_s}{V_p}$ $R_{ps} = f(A_s, V_p)$	$\frac{A_d}{V_p}$ $R_{pd} = f(A_d, V_p)$	$R_p = \sum R_{pj} / \sum R_{(pj)max}$
Total risk for each threat		$R_k = \sum (R_{ik} . P_i)$	$R_f = \sum (R_{if} . P_i)$	$R_s = \sum (R_{is} . P_i)$	$R_d = \sum (R_{id} . P_i)$	$Rt = \sum (Ri . Pi)$

Table 4.
Logical framework matrix for comprehensive risk assessment—eruption scenario.

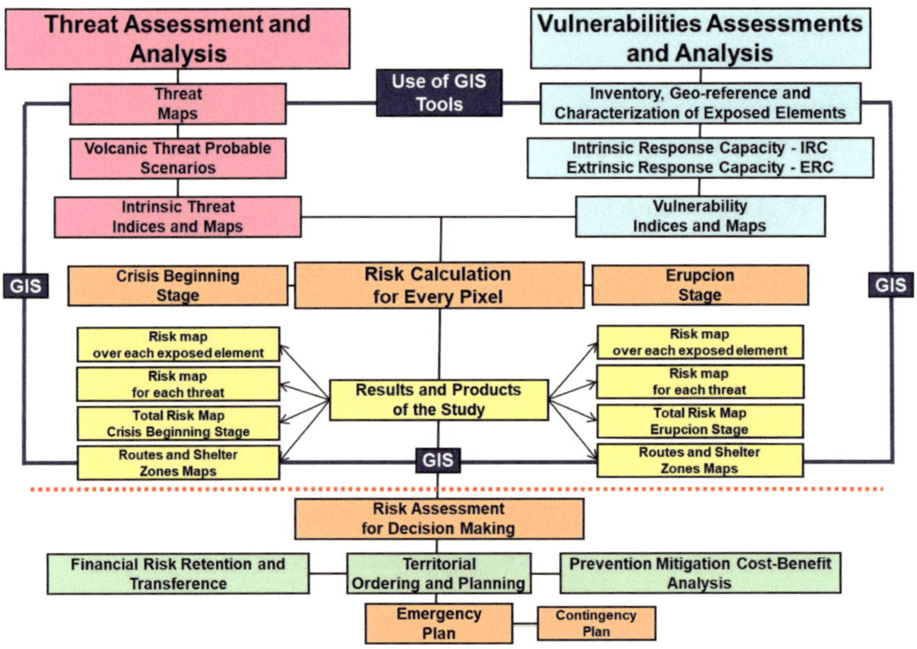

Figure 3.
Methodological diagram for the integral risk assessment.

of information; the alphanumeric and geospatial analysis with which it is possible to calculate the threat indexes, vulnerability, and risk at the level of each pixel of the study area; and finally the obtaining of products, such as risk maps for each analysis scenario.

What follows in this process diagram (after the dotted line in **Figure 3**) will be the risk assessment for decision-making in accordance with the retention and transfer of financial risk and with the cost-benefit analysis [24] that allows to define clear, precise, and consensual policy guidelines for land-use planning, as well as the corresponding emergency and contingency plans.

4. Case study: the Cerro Machín volcano

In the application of the conceptual and methodological approach previously exposed, the integral risk assessment is carried out in the Cerro Machín volcano, Colombia [16]. With this process, we obtain, for the scenarios of crisis initiation and eruption, the maps of total risk and the maps with the escape routes and zones of possible shelters for the transitory and/or definitive relocation of population and population centers, as shown below.

4.1 Total risk maps

According to the logical risk assessment framework, for each analysis scenario considered, the total risk maps are obtained by means of the weighted sum of the total risks on each exposed element, as shown in **Figure 4**.

According to the map in **Figure 4**, it can be seen in a general way that for the crisis initiation scenario, the highest total risk indexes are located in the areas near the volcanic building and in the valleys and slopes of the Toche and Bermellón rivers. For the eruption scenario, the high-risk indices are located in the areas exposed to the flows and landslides. The medium- to high-risk indexes are located in areas exposed to falls, characterized by the presence of crops and isolated rural housing.

4.2 Exit route maps and areas of possible hostels

The determination of escape routes and areas of possible temporary shelters and/or the final relocation of the population and population centers involves solving the following questions: When should evacuations take place? Who should be evacuated? Where and to where should they be evacuated? And so on. The answer

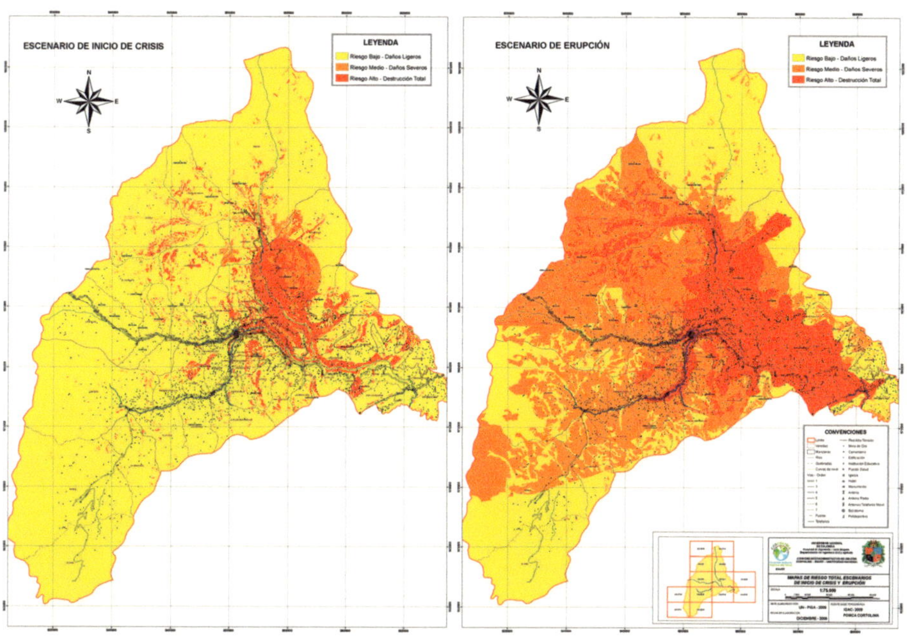

Figure 4.
Total risk duplex map—start of crisis and eruption scenarios.

to these questions constitutes a complex decision-making process that allows defining precise, clear, and consensual policy guidelines that guide, dynamize, and articulate a comprehensive risk management that adequately involves both the processes of comprehensive risk assessment and the formulation of emergency, contingency, and mitigation plans that guarantee efficient and effective risk management.

This process of making political, economic, administrative, logistical, social, environmental and technical decisions can be easily carried out if you have the right information and tools. This analysis should consider adjustments in land use planning, in the retention and transfer of financial risk, and of course, in cost-benefit analyzes for the prevention and mitigation of risk.

Taking into consideration the total risk maps for the two analyzed analysis scenarios, the maps are generated where the escape routes and areas of possible shelters for transient and/or definitive relocation of population and population centers are defined, as they are presented in duplex manner in **Figure 5**.

According to the above, some preliminary aspects to be considered in the process of progressive evacuation of risk areas are described below.

Pre-crisis scenario: according to [17], this scenario corresponds to the current situation of the volcano and may last from several to hundreds of years, in which the threats do not materialize. It is characterized by strong emanations of gases and by the eventual seismicity of the volcanic building that can cause some important landslides in the most susceptible areas and get to affect the nearest inhabitants and some communication routes. Due to the characteristics of this scenario and in light of the risk study carried out, one should now think of a definitive relocation strategy for the population settled in the high-risk area and begin immediately with the active participation of the community and the tasks of review, validation, and testing of emergency plans.

Start crisis scenario: according to [17], this scenario can last from hours to weeks prior to an imminent eruption and is characterized by the increase of seismicity, gas emissions, and the possible collapse of the south western side of the volcanic building, causing the damming of the Toche river and landslides in the areas near the

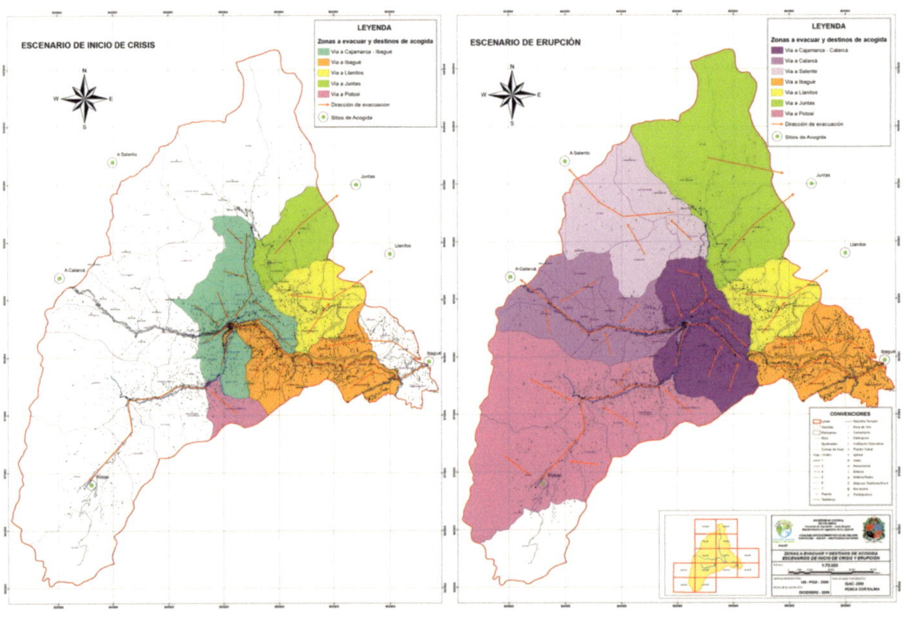

Figure 5.
Duplex map of escape routes and areas of possible hostels—scenarios for the initiation of crises and eruption.

volcano. According to the total risk map in the crisis start scenario, the lands that comprise the volcanic building, approximately five kilometers around the volcanic cone, are at high risk. Consequently, in this scenario, all efforts and contingency plans for the evacuation of people to save the greatest number of lives could still be safely implemented.

Eruption scenario: according to [17], this scenario can last from days to weeks, it includes the phases of minor eruptions, blast, and principal, and all threats are materialized, and the area of affectation is considerable. According to the total map of risks in eruption scenario (**Figure 5**), a large part of the study area that involves, among other elements, the population centers of Cajamarca, Anaime, Toche, Tapias, and Coello-Cocora, as well as the Pan-American Highway in the section between Ibagué and Cajamarca, are at high risk. Consequently, it would be expected that the occurrence of this scenario will ensure that the populations located in high risk areas have already been evacuated and relocated previously according to the map of escape routes and areas of possible hostels. Likewise, it would be expected that the populations located in areas of medium to low risk have already been prepared to start the evacuation processes to the recommended sites.

5. Conclusions and recommendations

The study establishes the bases of a new conceptual and methodological framework for the integral risk assessment, which, in addition to guiding the development of the study, allows the generation of information processing and analysis tools to make it possible.

According to the total risk maps for each analysis scenario, it is evident that in the event of an eruption of the Cerro Machín volcano, the population centers of Cajamarca, Anaime, Toche, Tapias, and Coello-Cocora, as well as the Pan-American Highway in the stretch between Ibagué and Cajamarca, would be seriously affected.

According to the map of escape routes and areas of possible hostels, it is ratified as an adequate site to relocate the populations of Cajamarca and Anaime to the Potosí sector, provided that it is complemented with the layout, design, and construction of a new road that communicates from Ibagué to Potosí and from there to Quindío, not to leave these populations isolated.

According to the map of total risks and in compliance with the precautionary principle, it is recommended to continue with the processes of education and preparation of the population for the emergency, indicating their escape routes and temporary and/or definitive shelter sites. It is considered pertinent to start already (immediately) the design of a comprehensive prevention strategy, which, on the one hand, orients the processes of relocation of the aforementioned population centers and, on the other, initiates the design, layout, and construction of a new route or route alternate that gives operational redundancy to the current Pan-American route.

It is recommended to convert the entire surrounding area to the Cerro Machín volcano in a large protected area attached to the national system of natural parks, which guarantees an ordering and territorial management more appropriate to the risks involved and its great ecotouristic potential. This option will allow, in principle, the reorientation of resources for the maintenance of access roads, and, in passing, the strengthening of risk management capacity. In the future, it will allow the development of low impact ecotourism activities, through hostels, thermal pools, restaurants, ecological trails, etc. as well as the construction of a cable car that facilitates rapid access from Cajamarca to the hill of San Lorenzo and from there to the Machín volcano crater.

It is evident that the integral evaluation of the risk is a determining factor in the processes of territorial ordering and therefore it is suggested the revision and adjustment to all the territorial planning plans that have not involved this factor, in a priority way in those watersheds and populated centers with obvious natural threats by volcanism, mass removal, torrential floods, avalanches, floods, forest fires, etc.

Finally, it is suggested to test and calibrate the model developed in the risk assessment of other volcanoes in the country.

Acknowledgements

To the Universidad Nacional de Colombia, to the Faculty of Engineering, to the Extension and Research Institute, to the Department of Civil and Agricultural Engineering for the unconditional support. To Dra. Carmen Sofía Bonilla, General Director of the Corporación Autónoma Regional del Tolima (CORTOLIMA), who for her integrating vision between environmental management and risk management made it possible to institutionally materialize the development of the study through the Universidad Nacional de Colombia. To CORTOLIMA, for its decisive technical and logistical support in providing the thematic information for the development of the study and in particular to Dr. Ulises Guzmán, Interventor on behalf of CORTOLIMA, for his accurate and timely observations, suggestions, and recommendations. To the Ministerio de Ambiente, Vivienda y Desarrollo Territorial (MAVDT) for its financial support to defray part of the research costs. Finally, to the entire professional and academic team of the PIGA Group for Research in Politics, Information, and Environmental Management but in a special way to the Civil Engineer, MSc. Geotecnia Fernando Javier Díaz, for his determined collaboration and valuable contributions, particularly in relation to the evaluation of threats due to seismicity and landslides and geospatial analysis.

Author details

Leonel Vega
Universidad Nacional de Colombia, Bogotá, Colombia

*Address all correspondence to: lvegamora@unal.edu.co

References

[1] Blong RJ. Volcanic Hazards: A Source book on the Effects of Eruptions. Sydney: Academic Press; 1984

[2] Cuellar J, Ramírez C. Descripción de los volcanes colombianos, In: Primer Simposio Internacional sobre Neotectónica y Riesgo Volcánico, Bogotá; 1986

[3] Méndez R. Catálogo de los volcanes activos de Colombia. Boletín Geológico. 1989;**30**(3):1-75. Ingeominas

[4] Tilling RI. Lessons in reducing volcano risk. Nature. 1993;**364**:277-280

[5] Pomonis A, Spence R, Baxter P. Risk assesment of residencial buildings for an eruption of Furnas Volcano, Sao Miguel, The Azores. Journal of Volcanology and Geothermal Research. 1999;**92**:107-131

[6] Domínguez L, Concha A, Cepeda H. Volcán Cerro Machín: Mecanismo eruptivo, Transporte y Sedimentación (Tolima, Colombia). Geología Colombiana No. 28. Bogotá; 2003

[7] Thuret JC. Urban hazard and risk; consequences of earthquake and volcanic eruptions an introduction. GeoJournal. 1999;**49**:131-135

[8] Ochiai H, Yanase H, Matsuura S, Takeuchi Y. Land slide initiation mechanism of Bandai volcano effect of earthquake motion. Bandai Volcano. Science and Technology Agency. Japan, 1997: 273-283

[9] Obando M, Ramos R, Portilla M, Cepeda H. Modelamiento asistido por computador de flujos piroclásticos en el volcán Cerro Machín (Tolima, Colombia). Geología Colombiana No. 28; 2003

[10] CAPRA – ERN. Modelos de Evaluación de Amenazas. 2011. Available from: www.ecapra.org

[11] Cerisola C. Resiliencia y Programas Preventivos. El Salvador: Universidad del Salvador, Facultad de Psicología; 2003

[12] Chamochumbi W. La Resiliencia en el Desarrollo Sostenible. 2005. Available from: Ecoportal.Net

[13] LA RED. Ciudades en Riesgo, Degradación Ambiental, Riesgos Urbanos y desastres. Red de Estudios Sociales en Prevención de Desastres en América Latina. 2006. Available from: www.researchgate.net/scientific-contributions/47148707

[14] OSSO. Estudio de Vulnerabilidad Física y Funcional a Fenómenos Volcánicos en el Área de Influencia del Volcán Galeras. Available from: http://osso.org.co:8000.2009

[15] Vega L. Toward the systemic parameterization of the environmental dimension. Revista Ingeniería e Investigación. 2011;**31**(1):242-253. ISSN: 0120-5609

[16] UN-PIGA. Estudios y análisis de vulnerabilidad y riesgo en un sector del área de influencia del volcán Cerro Machín. Informe final Convenio Interadministrativo UN-CORTOLIMA-MAVDT No. 088 de 2008. Universidad Nacional de Colombia. Grupo PIGA de Investigación en Política, Información y Gestión Ambiental. Bogotá; 2009

[17] INGEOMINAS. Evaluación de la Amenaza Volcánica Potencial del Cerro Machín. Manizales, Colombia: Departamento del Tolima; 2002

[18] Cardona O. Estimación holística del riesgo sísmico utilizando sistemas dinámicos complejos [Tesis doctoral]. Barcelona, España: Universidad Politécnica de Cataluña; 2001

[19] Cardona O. Midiendo lo Inmedible. Indicadores de Vulnerabilidad y Riesgo.

2005. Available from: http://www.desenredando.org/public/omar.html

[20] Vega L. La Dimensión Ambiental del Desarrollo. ECOE Ediciones. Colección Ciencias Naturales. Bogotá: Área Ecología y Medio Ambiente; 2017. ISBN: 978-958-771-390-9 (papel). ISBN: 978-958-771-425-8 (digital)

[21] Wilches-Chaux G. Fundamentos éticos de la gestión del riesgo. Nómadas (Col). No. 22, Bogotá; April 2005. pp. 48-61. ISSN: 0121-7550 (printed), ISSN: 2539-4762 (online)

[22] USAID. Curso Básico de Sistema de Comando de Incidentes (SCI). Manual de Participantes. USA. 2005. Available from: https://scms.usaid.gov/sites/default/files/documents/1866/Material%20de%20Referencial%20SCI.pdf

[23] IGAC. Tutorial Sobre Fundamentos de Sistemas de Información Geográfica. 2004. Available from: www.igac.gov.co

[24] Cardona O. Evaluación del Riesgo de Desastre con Fines de Estimación de Pasivos Contingentes y Déficit Fiscal. 2006. Available from: http://www.desenredando.org/public/omar.html

Chapter 6

The Somma-Vesuvius Activity with a Focus to the AD 79 Eruption: Hazard and Risk

Angelo Paone

Abstract

Somma-Vesuvius is a quiescent stratovolcano with a probability of Plinian style volcanic reactivation. Its stratigraphy is well known in the last 40 ka BP. The volcanic products that are part of the Somma caldera are poorly studied. Conversely, younger products have been deeply studied together with the AD 79 Plinian eruption. The impact of a Plinian eruption has been studied and summarised here. A simplified scheme is presented from what we can understand the volcanic hazard and risk that the volcano poses to the greater Neapolitan population. In the last 40 years, the demography around the Somma-Vesuvius volcano has increased; consequently, the volcanic risk has increased. It would seem that the Italian Civil Protection (ICP) has not influenced the population and the Italian authority with their massive work around Somma-Vesuvius (red zone). People still continue to build houses. Nowadays, the Somma-Vesuvius volcano does not seem to threaten people or the people that live around Vesuvius are not afraid of the volcano. But as it is usual just in this moment that the work done and to be done must be speared to all Neapolitan people, working in the school to reach the family. People around Somma-Vesuvius tend to neglect the volcanic risk appearing around Vesuvio. So ICP, all must be much more attend about the behaviour of this Hazardous volcano.

Keywords: Somma-Vesuvius, AD 79 eruption, Plinian, Hazard, Pompeii

1. Introduction

Somma-Vesuvius volcano has shaded many volcanological avenues since the two letters of Pliny the Younger to Tacitus [1]. He described the dynamism of the Pompeii Plinian eruption. Since then, Somma-Vesuvius has utterly been studied by many earth scientists worldwide who have increased the knowledge of the volcano. Today, Somma-Vesuvius is the best studied and monitored volcano in the world. The volcano is very much known about its volcano stratigraphy and by the chemistry of the volcanic products as well as by the physical architecture of the volcano. Anyhow, it is little known how the population feel about the impact of the hazardous volcanic eruptions [2–6]. To envisage different volcanic scenarios is a must in order to help the population around Vesuvius to better understand the type of next eruption that will impact on the region nearby Somma-Vesuvius. A simplified scheme of AD 79 eruption related to all the volcanic activity of a such destructive volcano can shed

IntechOpen

light and ease the Neapolitan population that live around Somma-Vesuvius in case of future Plinian eruption. This simplified review on an open access international journal can also be available for students and general population and help high school students to understand the hazard of Somma-Vesuvius and to discuss this issue with their family.

2. Short notes on the stratigraphy of Somma-Vesuvius in the last 40 ka BP

The volcanic stratigraphy has been studied in detail in the last 40 ka BP [7]. Before this age, the volcano was looking like the Fuji volcano in Japan today (**Figure 1**). This analogy can be a best guess of the Somma volcano before 40 ka BP. The eruptive history of Somma volcano can be studied only on the Somma Caldera wall, which is poorly studied. Today, the volcano called Somma-Vesuvius volcano looks like the one shown in **Figure 2**. The aerial photo shows clearly the two morphologic characteristics: Somma caldera and Vesuvius cone nested into the

Figure 1.
Fujiyama volcano, Japan (taken from the Web).

Figure 2.
Aerial photo of Somma-Vesuvius (Somma caldera e Vesuvius cone).

caldera. The Somma caldera is well exposed and looking from the North side (from Pollena town); the Vesuvius nested cone is hidden (**Figure 3**). Somma volcano should have been around 2000–2500 m high extrapolating from Somma Caldera, much higher than actual altitude of Vesuvius cone (Rolandi, unpublished data) [8]. The analysis of the eruptive history of Somma-Vesuvius (**Figure 4**) reveals the recurrence of highly explosive events followed by repose periods, during some of which low (2-3-4) VEI strombolian and/or vulcanian style eruptions took place [9]. The last eruptive cycle at Vesuvius lasted from 1631 to 1944 [9]. The AD 1944 eruption is the last eruption of Vesuvius Cone. Now, the volcano is in a repose period, which can be followed by a Plinian eruption. The behavior of Vesuvius volcano in the last 3800 years BP can be seen in **Figure 5** [9]. After the famous Avellino

Figure 3.
Somma caldera seen from North to West.

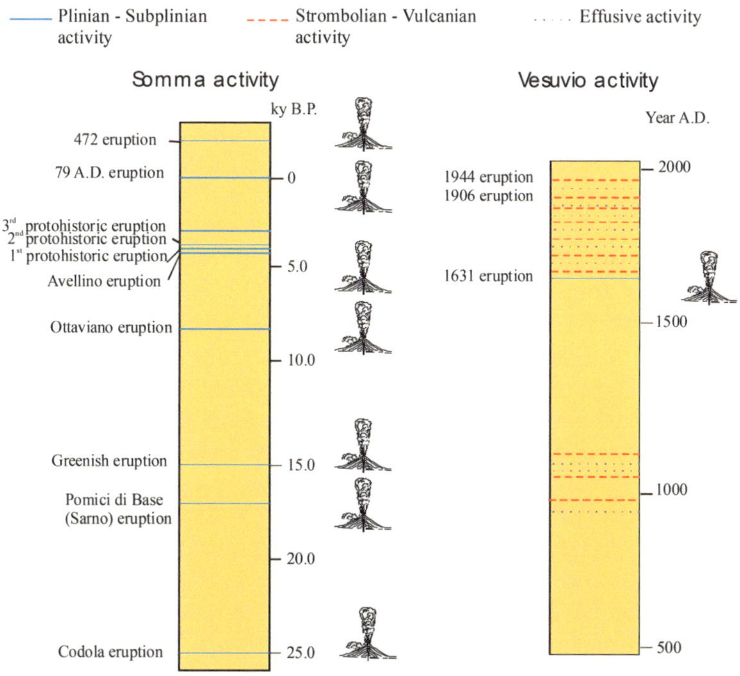

Figure 4.
Somma-Vesuvius stratigraphic sketch with Somma activity and Vesuvius activity with style of eruption.

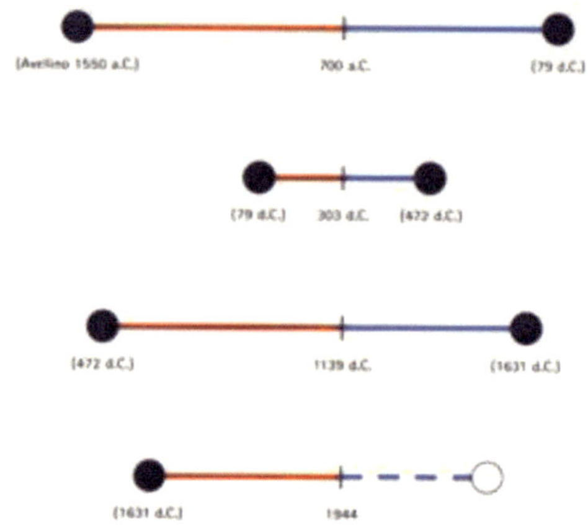

Figure 5.
Activity of Somma-Vesuvius from Avellino eruption to the 1944 eruption: Plinian, interplinian and repose time [9]. The interplinian is defined as volcanic cyclic period less explosive after each Plinian eruption.

prehistoric Plinian eruption, the Somma changed its eruption behavior, and the sequence of Plinian-interplinian-repose time took place until 1944. The protohistoric interplinian phase was followed by 700 years of repose time before the AD 79 eruption (**Figures 4** and **5**) [9]. According to this sequence, the next eruption at Vesuvius would be of the Plinian style. The Vesuvius matter would be to guess the size of the next Plinian eruption: Avellino-like, AD 79 eruption-like, Pollena-like, and 1631 eruption-like. This is still matter of study if the future eruption will be Subplinian, Plinian, or Ultraplinian.

3. Somma volcano before the AD 79 eruption

Somma volcano, *Vesuvio* for Roman people, before the AD 79 eruption, raised no worry to people living in the volcanic area. The wall painting found in the archaeological excavation of Pompeii depicts the volcano as a one-peaked mountain (**Figure 6**). Bacco indicates the pleasant life of the time and the fertility of soils. No reference is made to the presence of phenomena (e.g., fumaroles) typical of an active volcanic area. Even in Latin literature (Virgilio, Seneca, Pliny the elder, Columella), the volcano is known only for fertility of its soils. Vitruvio is the only writer to put forward the hypothesis that in the past, although all his contemporaries had forgotten the fact, the volcano had been devastated by powerful eruptions (**Figure 6**). The Greeks wrote of the age, on the contrary, as Diodoro Siculo (80–20 BC) and the geographer Strabo (1st century BC–AD 19) describe morphological features of Vesuvio volcano: *...flat, without vegetation, similar to ash in appearance, with porous holes in the rocks which are black in colour as if they had been burnt... This is probably why the soils are so fertile ... like Catania... covered by the eruptive ashes from Etna's fire...* (Strabo) (**Figure 7**). Even seismic activity was not taken into account as a precursor of an eruption, but it was more simply linked to seismogenetic activity of the Apennine Chain. Seneca, in Quaestiones Naturales, does not quote Vesuvio when he describes the Pompeii earthquake on February 5 AD 62. The author, not aware of the seismogenesis of volcanic areas, does not recognize this earthquake as a possible precursor of a volcanic event (**Figure 8**). In [8, 10], it is shown how the evolution of the Somma caldera has been very complex with only the post activity of AD 472 called interplinian started with the growth of Vesuvius cone [9].

Figure 6.
Drawing of Somma volcano during the AD 79 eruption (note a one peaked volcano conversely to the actual two-peaked volcano). Artist unknown found in one Pompei room.

Figure 7.
Drawing of Somma volcano after the Codola eruption and probably Sarno eruption.

Figure 8.
Roman marble sculpture witnessing the earthquake of the AD 62.

4. Date of the AD 79 eruption

The south-easterly trend of the AD 79 products appears to be anomalous, because the eruption is conventionally believed to have occurred on August 24, when its southeast dispersive trend falls in a transitional period from the summer to autumnal wind regimes [11]. In fact, the AD 79 tephra dispersive direction toward the southeast is not in agreement with the June–August high-altitude wind directions in the region that are rather toward the west. This poses serious doubt about the date of the eruption and the mismatch raises the hypothesis that the eruption occurred in the Autumnal climatic period (October), when high-altitude winds were also blowing toward the southeast. New archaeological findings presented in the [11] study definitively place the date of eruption in the Autumn (October), in good agreement with the prevailing high-altitude wind directions above Somma-Vesuvius ([11]; references therein).

5. The chronology of AD 79 eruption: How the eruption impacted on the Roman population

Pliny the Younger observed the eruptive column from Miseno at about 13 h of the day of October 24 (**Figure 9**) [1, 11]. Probably, the start of the AD 79 Pompeii eruption (**Figure 10**) and the eruptive column begin to rise probably at 12 h of October 24 and appear just as strange phenomena for the Pompeii people (**Figure 11**). Fall-out products from the AD 79 eruption can be found both in the surrounding area and at a distance, stretching in a SE direction as far as Cilento: October 24 (the first day of eruption): "...a cloud appeared of unusual size and shape... The cloud advanced in height; and I cannot give you a more just representation than the form of a pine tree..." [1]. Eruptive column of the white pumice was sustained for about 8 h. Pliny the Younger, in its letters to the historian Tacito,

Figure 9.
The younger Pliny reproved.

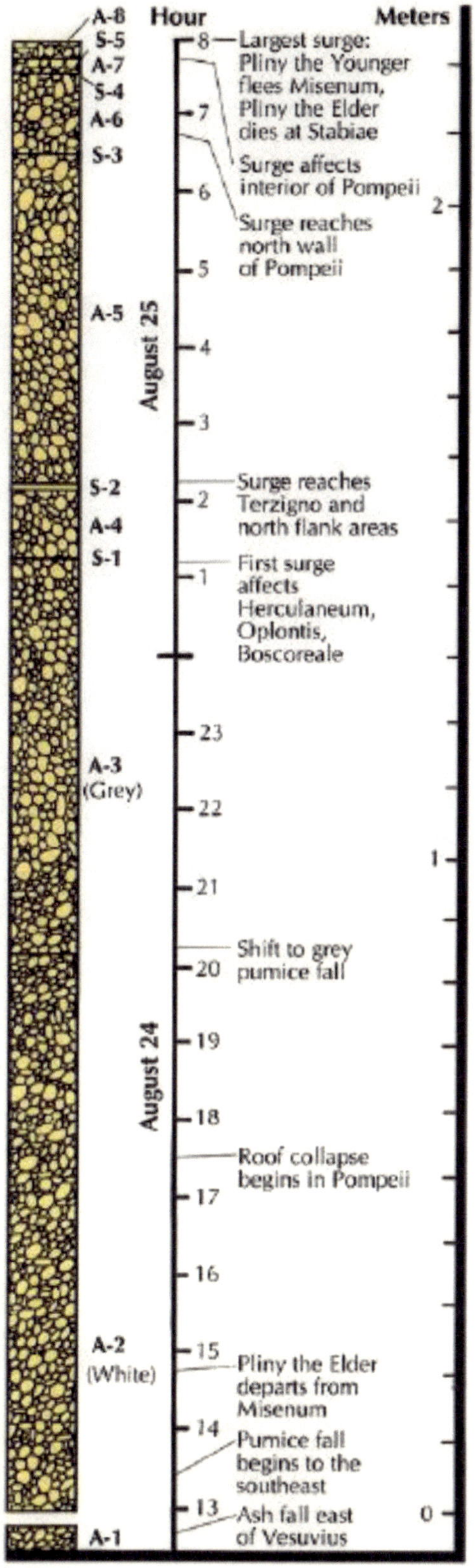

Figure 10.
Modified volcanic section of the AD 79 [20].

Figure 11.
Cartoon shows how the Romans were surprised at the begin of the AD 79 eruption. They were unaware of the volcanic hazard.

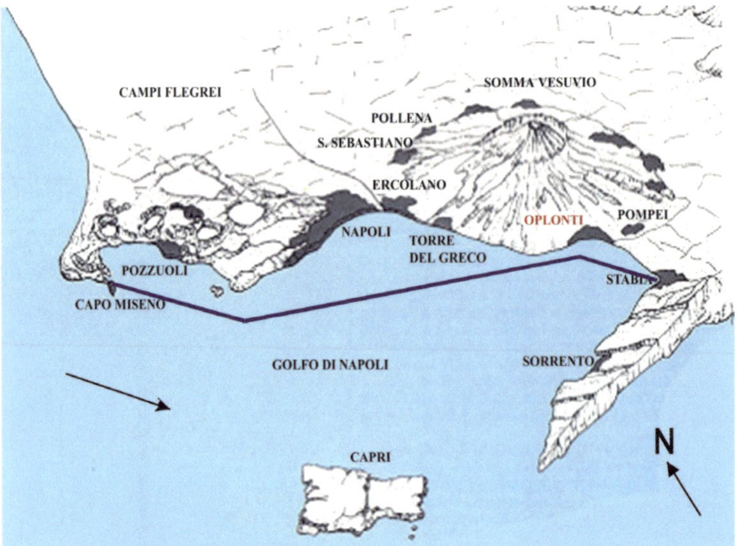

Figure 12.
The route of Pliny the Elder; after that, he realized the start of Somma activity (AD 79 eruption).

writes that his uncle Pliny the Elder "... was at Misenum where he had the command of a fleet which was stationed there. At about 15 h Pliny the elder received a request of help by his friend Rectina and decided to start from Misenum with four Liburnae. After 4 h, he arrived near the coast of Oplonti but he cannot disembark because the effects of the eruption and decided to move toward Stabie, where his friend Pomponiano was located (**Figure 12**). At Pompeii, roof collapse due to the weight of the white pumice level begins between 17 and 18 h (**Figure 13**). After that, the Pompeii eruption changed from white pumice to grey pumice. This transition also changed the composition of the eruption (**Figure 14**). During the first day of the eruption Herculaneum was spared by pumice fall phase and people could observe the pyroclastic cloud diverted toward SE by winds. From about 1 to 8 h of

Figure 13.
Cartoon shows the moment when Pompeii home start to collapse by the load of fall-out volcanic phase.

Figure 14.
Classic section of the AD 79 eruption with white and grey pumice fall-out.

October 25, the Plinian column of grey pumice assumed a collapsing character giving rise to surge and pyroclastic flows currents. Pliny the younger wrote I believe, while the vapor was fresh, it (the cloud) more easily ascended; but when the vapor

Figure 15.
Herculaneum with Roman community hit by pyroclastic density currents at the end of white and white and grey pumice fall-out.

was wasted the cloud became loose, or, perhaps, oppressed by its own gravity, and dilated itself into a greater breath..."

October 25 (the second day of eruption): pyroclastic flows bury the town (Herculaneum) under a 15-m-thick ash deposit. The inhabitants, in the vain attempt to escape by boat, take refuge in the barrel-vaults along the seaside where they will die (**Figure 15**). The Pompeii and Herculaneum towns, before the eruption, nearly looked out directly over the sea. The AD 79 eruption was a natural event that strongly affected the perivolcanic area of Somma-Vesuvio. The paleogeographic and socioeconomic variations that it caused were huge. After the eruption, the area was completely buried with the consequent progradation of the coast line [12]. Trading in the Nola and Acerra areas was badly affected as communication routes to the sea through the port of Pompeii no longer existed. Toward Stabia, the Pomponiano house was highly damaged by white and grey pumice fall, and the courtyard was filled by pumice deposits, so that early in morning of October 25, Pliny the elder decided to move toward the beach. In the meantime, a phase of eruptive calm had been registered and, in the Pompeii area, people who had managed to escape the

Figure 16.
Cartoon shows how the PDC were severe by hitting Roman villages.

Figure 17.
Engraving showing Pliny the Elder found death on the beach of Stabie village.

effects of the pumice fall phase tried to return to their homes by walking over the pumice deposits which had become as high as the rooftops. As they made their way home, accompanied by some soldiers, hot toxic cloud swamps departed by the hot avalanche that are invading Herculaneum, and kills them (**Figures 16** and **17**).

October 26 (the day after the eruption): Pliny the Elder's body was found dead on the beach "...when the light returned, which was not till the third day after his death, his body was found untouched by the fire, without any visible hurt, in the dress in which he fell, appearing rather like a person sleeping than like one who was dead" (Pliny's letters).

The destructive action of pyroclastic products on constructions is exerted in two main ways. Fall products, which are emplaced grain by grain, falling down from the eruptive cloud, blanket roofs with a layer of granular deposit, whose thickness is a function of both the intensity of the eruption and the distance from the vent. The pyroclastic fall layer loads the roofs with an extra weight that, when the maximum sustainable load threshold value is exceeded, can cause the roof to collapse. Pyroclastic density currents (PDC), on the other hand, generally originating from the collapse of the eruptive column, are gas clouds rich in both lithic and juvenile fragments that, running over the ground, exert a dynamic overpressure on the obstacles encountered on their way and are capable to destroy the buildings (**Figure 18**) [13]. Actually, the first damage was made by white pumice fall deposits,

Figure 18.
Effects of AD 79 eruption on Roman buildings: Villa 6 at Terzigno.

Figure 19.
Romans' body-mould made by gypsum in modern age.

weighing on the roofs, leading to collapse them. The first pyroclastic density currents occurred during the late deposition of white pumice fall products; successively, grey pumice fall and pyroclastic flow and then PDC. The broken columns of the main atrium lie on the pumice fall deposits. The collapsing angle of the columns points out also the direction of the diluted currents (**Figure 18**). Around Somma-Vesuvius volcano, there is also a petrified tree that by its curvature shows the direction of the PDC. The first day of eruption produced 3 m of pyroclastic fell products blanket Pompeii, and some inhabitants fell asleep, and were buried by the collapse of their houses. Steps needed to obtain a mould of a body buried by pumice fall are: (1) the body was covered by pumice fall during the eruption; (2) organic matter was destroyed by heat, and the vacuum had been filled later on, in modern age, by gypsum; and (3) the shape of the body was then recovered (**Figure 19**).

6. The Somma-Vesuvius history after the AD 79 eruption

The history of ancient stratovolcano that suffered a caldera-like collapse, Somma [8, 10], ended with the AD 472 eruption. In Middle Ages, from AD 500 to AD 1100, the mediaeval interplinian period occurred. During this period, the volcano alternates quiescence periods to effusive phases with low energy Strombolian style eruptions. The last reliable historical reports about this long time span concern the effusive activity that took place in 1139. The mediaeval interplinian activity was responsible

Figure 20.
Engraving showing Somma-Vesuvius during a repose period before the 1631 eruption.

Figure 21.
Portici marble label as the first action of Civil Protection at Somma-Vesuvius after the 1631 eruption.

for the construction of a new stratovolcano, the Vesuvio, which grew within the Somma caldera [14–19]. Since 1550, many coeval chronicles and the pictures of the Atlas by the cartographer Gerardus Mercator (1512–1594) make it possible to deduce that in this time span the Vesuvius was quiescent and that Gran Cono was covered by a closely cropping vegetation. On December 16, 1631, Vesuvius resumed its activity after 500 years of quiescence period. The eruption, with both effusive and explosive products, was not expected by the inhabitants, who were not aware that Vesuvius was an active volcano (**Figure 20**). Famous label was placed at Portici by Spanish viceroy after the AD 1631 eruption to warn the future generation about the volcanic risk of the Vesuvius (**Figure 21**). From 1631 to 1944, the last interplinian phase occurred. The typical volcanic activity, effusive and weak explosive, was developed according to a succession of volcanic cycle, called vesuvian cycle. Between the end of eighteenth century and the first half of the nineteenth century, in the absence of photographs,

a

b

c

Figure 22.
(a) December 1760 eruption, Fabris. (b) October 20, 1767, Fabris (c) June 1794 eruption.

grew, in Napoli, painters "a la gouaches." These artists represented the wonderful landscapes of the vesuvian area. Their pictures are the precious examples of the Neapolitan "vedutismo" (**Figure 22**).

A progressively increasing of both strombolian and effusive intracrateric activity preceded the 1822 paroxysm. In September, an appreciable seismic activity was recorded and also many springs of the vesuvian area dried up. After some volcanic eruptions, on October 20 and 21 during the night, finally the lava overflowed the crater rim and directed toward Resina (Ercolano) (**Figure 23**). In these years, on the southern flank of Vesuvius, above the Salvatore hill, the building of the first volcanological observatory of the world, the **Osservatorio Meteorologico Vesuviano**, was built. It was officially inaugurated on September 28, 1845, during the Conference of Italian Scientists held in Napoli.

This particular moment in history signaled for the people living there at a transition from a phase of fear of the volcano to one of lasting and increasing knowledge (**Figure 24**). The 1906 eruption was the paroxysmal phase of the Vesuvian cycle

Figure 23.
Interplinian eruption of 1822.

Figure 24.
Historical Vesuvian Observatory (first volcanological observatory on earth).

1874–1906. The eruption was observed and described by many Italian and foreign volcanologists. For the first time with 1906 eruption, Vesuvius photos were diffused in the scientific world also due to the contribution of newspapers and magazines (**Figure 25**). In 1913, the last Vesuvian cycle of the modern historic interplinian phase started, and on March 18, 1944, the paroxysmal eruption began (**Figure 26**). It was not long after the war that Italian people, especially those around Vesuvius, were busy sorting out other problems. The volcano eruption almost seemed like the lesser of two evils when compared to the destruction wrought by the Second World War in Campania. During the eruption, the Allied Force Command was involved in bringing an enormous help, through evacuation, to the population of Vesuvian municipalities and, in particular, of San Sebastino, which was firstly threatened and then invaded by lava flows (**Figure 27**). In 1944, Vesuvius entered a quiescence phase still now lasting, even if every now and then low magnitude seismic events,

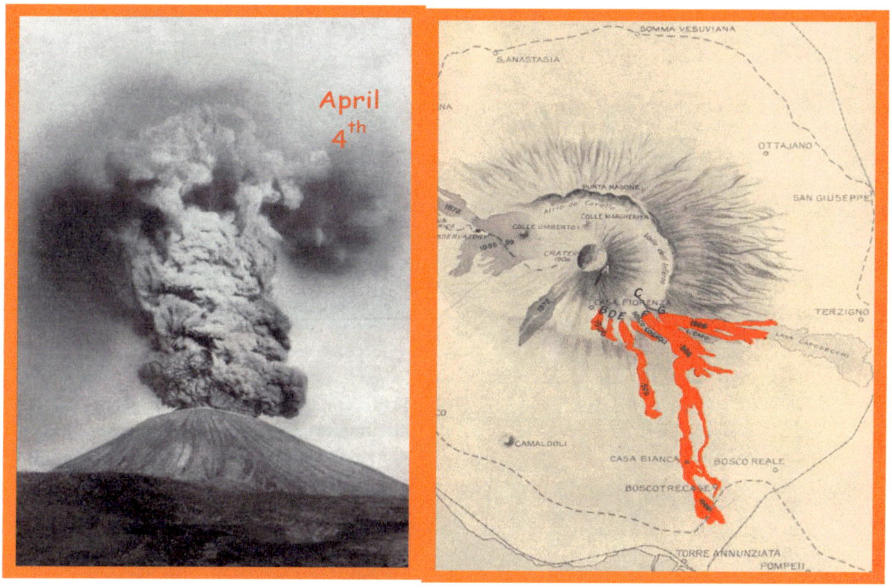

Figure 25.
1906 eruption: eruptive column and lava flow.

Figure 26.
1944 eruption: beginning of the eruption followed by a Vulcanian style eruption with an allied plane that suggest the height of the 1944 eruption column.

Figure 27.
1944 lava flow invaded the San Sebastiano al Vesuvio.

Figure 28.
Vesuvius crater after the 1944 explosive eruption.

together with a continuous fumarolic activity at the crater, remind people that, although dormant, the volcano is still active (**Figure 28**).

7. Estimating hazard and risk at Somma-Vesuvius

The last eruption at Vesuvius took place in 1944. After this date, the long lasting quiescence period encouraged the strong urbanization in perivolcanic areas. In the first decades of the twentieth century, when activity at Vesuvius was almost continuous, no significant increase in population rate was recorded along the Vesuvius coastline (**Figure 29**). In the last five decades, on the contrary, a very strong increase in population rate was recorded. In the coastal

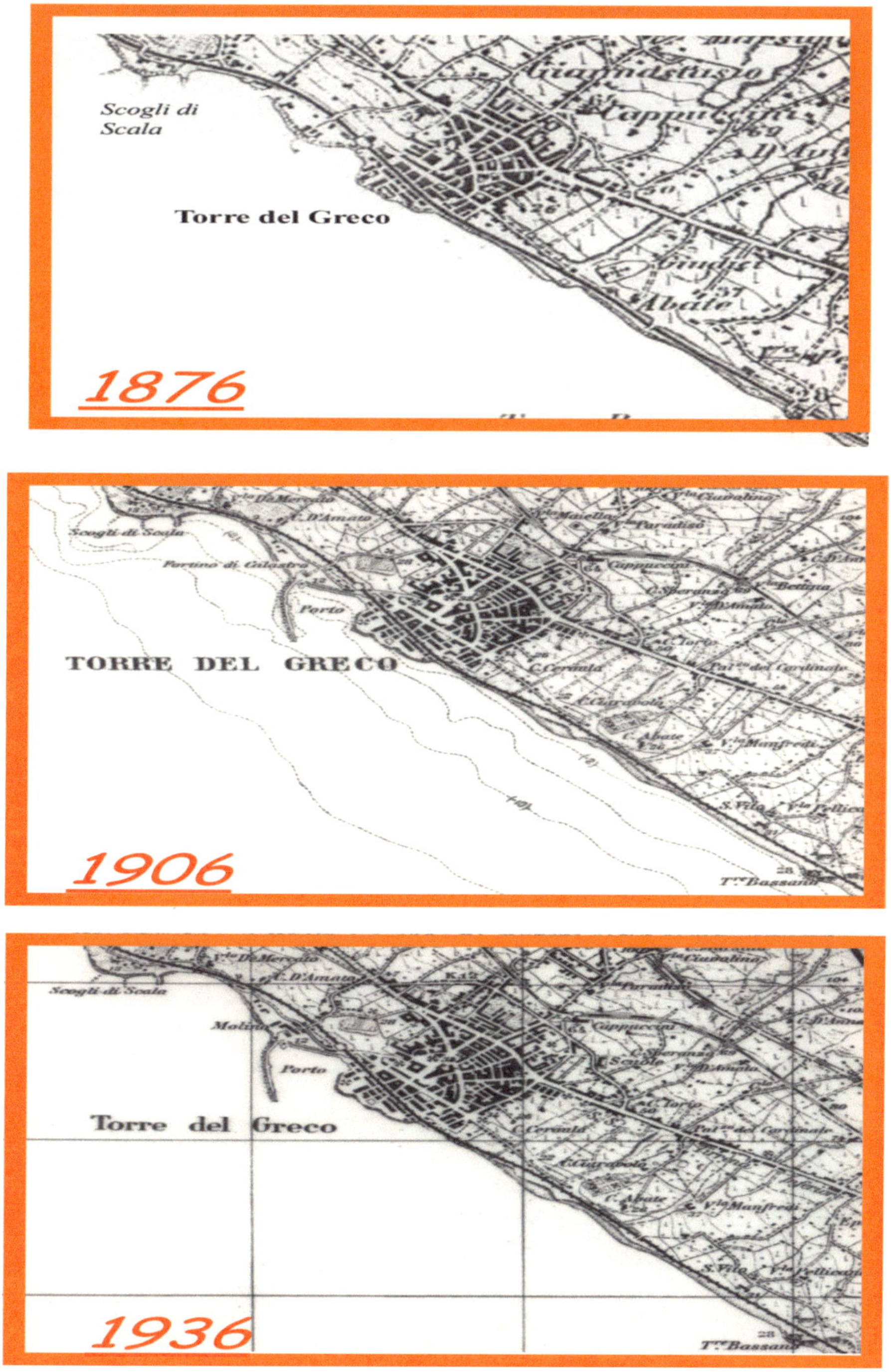

Figure 29.
Topographic maps of Torre del Greco by the following age: 1876, 1906, and 1936. This range does not represent an increase of the inhabitants. Data are taken from Italian newspapers.

area of Vesuvius, the population strongly increased up to 1980, when the large towns, already congested by traffic and degradation, become saturated (**Figure 30**). In 1930, inhabited areas were far from each other and located along the main municipal roads. In 1960, inhabited areas expanded and industrial sites grew in agricultural areas. At present, both inhabited and industrial

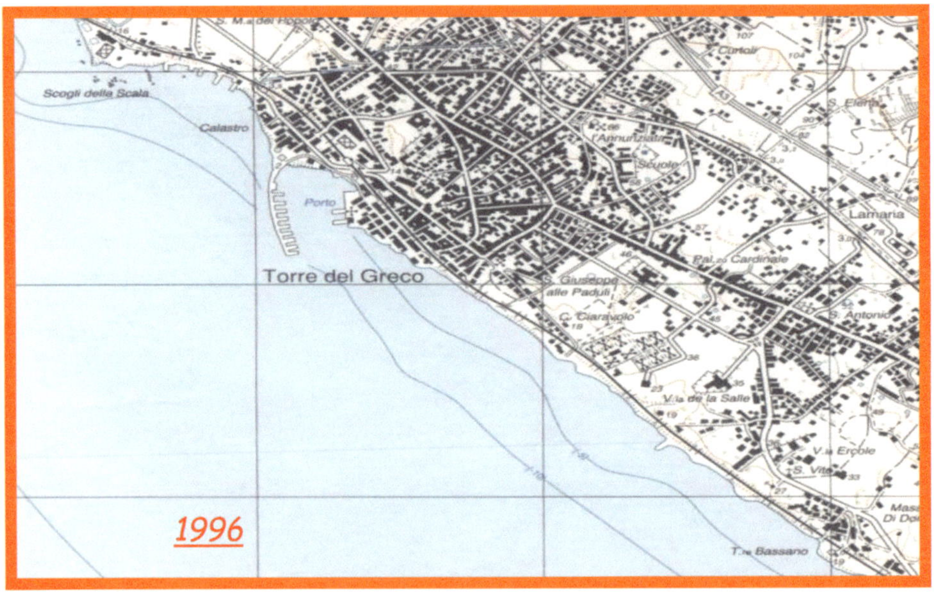

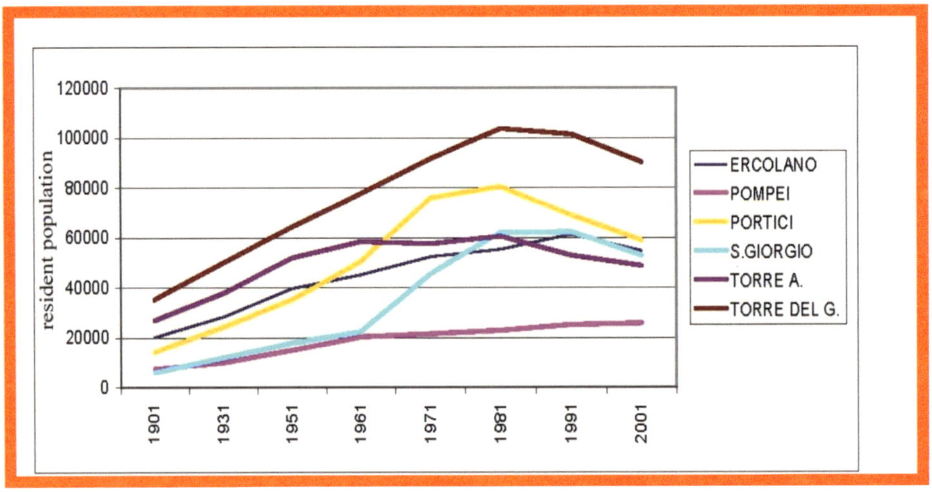

Figure 30.
1996 topographic map and a diagram pointing the increase of population in recent years (1996, unpublished data). From 1996 up to now, the increase of population has not stopped.

sites cover the whole territory, and agricultural areas are much reduced. A hazard map was obtained encompassing both the fall products possible hazard and the pyroclastic density current areas (drawn after Protezione Civile Italiana) (**Figure 31**).

8. Conclusion

The Somma-Vesuvius volcano is a composite stratovolcano with a complex history of caldera evolution. A sort of stratigraphic cyclicity is envisaged which suggests the style of the next eruption as Plinian. Anyhow, on the basis of size, it is very difficult to select an "ad-hoc" Plinian eruption which could be selected between Avellino, AD 79, AD 472, and AD 1631 eruptions. There is still a lot of

Figure 31.
Volcanic hazard and risk at Somma-Vesuvius.

confusion on this issue. Hazard and risk are very high due to the demographic increase around Vesuvius in the last 40 years (**Figure 30**). One of the last research [11] suggests that Napoli Municipality could be hit by a Plinian eruption so that hazard and risk would increase. From the volcano history, it can be noted that people have always forgotten the hazard and risk that Vesuvius pose. On the other hand, the Spanish have warned the population (1631 eruption) of the Hazard of Vesuvius Portici marble label (**Figure 21**). This protective action together with the construction of the Vesuvian Observatory indicates how Spanish were sensible about the behavior of the volcano. The AD 79 has been very well studied so that would be easy (cartoon model) for students understand the volcanic hazard. The volcanic section of the AD 79 drawn by Sigurdsson et al. [20] is still valid as a model to present how the AD 79 eruption has evolved and in parallel how Pliny the Younger wrote his letter.

Acknowledgements

Harvey Belkin (USGS) is thanked for looking to the English style. Karoly Nemeth is thanked for a pretty and thoughtful review of the manuscript.

Author details

Angelo Paone
Department of Earth Science Education, Pusan National University, Busan, Republic of Korea

*Address all correspondence to: angelo.paone1@gmail.com

References

[1] Lobley Logan J. Mount Vesuvius. A Descriptive, Historical, and Geological Account of the Volcano and Its Surroundings. London: Roper & Rowley; 1889

[2] Zuccaro G, Cacace F, Spence RJS, Baxter PJ. Impact of explosive eruption scenarios at Vesuvius. Journal of Volcanology and Geothermal Research. 2008;**178**(3):416-453

[3] Deligne NI, Fitzgerald RH, Blake DM, Davies AJ, Hayes JL, Stewart C, et al. Investigating the consequences of urban volcanism using a scenario approach I: Development and application of a hypothetical eruption in the Auckland volcanic field, New Zealand. Journal of Volcanology and Geothermal Research. 2017;**336**:192-208

[4] Blake DM, Deligne NI, Wilson TM, Lindsay JM, Woods R. Investigating the consequences of urban volcanism using a scenario approach. II: Insights into transportation network damage and functionality. Journal of Volcanology and Geothermal Research. 2017;**340**:92-116

[5] Sieron K, Ferres D, Siebe C, Constantinescu R, Capra L, Connor C, et al. Ceboruco hazard map. Part II: Modeling volcanic phenomena and construction of the general hazard map. Natural Hazards. 2019;**96**(2):893-933

[6] Kereszturi G, Nemeth K, Cronin SJ, Procter J, Agustin-Flores J. Influences on the variability of eruption sequences and style transitions in the Auckland volcanic field, New Zealand. Journal of Volcanology and Geothermal Research. 2014;**286**:101-115

[7] De Vivo B, Rolandi G, Gans PB, Calvert A, Bohrson WA, Spera FJ, et al. New constraints on the pyroclastic eruptive history of the Campanian volcanic plain (Italy). Mineralogy and Petrology. 2001;**73**:47-65. DOI: 10.1007/s007100170010

[8] Rolandi G, Bellucci F, Cortini M. A new model for the formation of the Somma caldera. Mineralogy and Petrology. 2004;**80**:27-44

[9] Rolandi G, Petrosino P, McGeehin J. The interplinian activity at Somma-Vesuvius in the last 3500 years. Journal of Volcanology and Geothermal Research. 1998;**82**:19-52

[10] Cioni R, Santacroce R, Sbrana A. Pyroclastic deposits as a guide for reconstructing the multi-stage evolution of the Somma-Vesuvius caldera. Bulletin of Volcanology. 1999;**60**:207-222

[11] Rolandi G, Paone A, Di Lascio M, Stefani G. The 79 AD eruption of Somma: The relationship between the date of the eruption and the southeast tephra dispersion. Journal of Volcanology and Geothermal Research. 2007;**169**:87-98

[12] Pescatore T, Senatore MR, Capretto G, et al. Holocene coastal environments near Pompeii before the A.D. 79 eruption of Mount Vesuvius, Italy. Quaternary Research. 2001;**55**(1):77-85

[13] Branney MJ, Kokelaar P. Pyroclastic density currents and the sedimentation of ignimbrites. Geological Society of London's Memoirs. 2003;**27**:1-143

[14] Cioni R, Isaia R, Sulpizio R, de Vita S, Di Vito MA, Pistolesi M, et al. The city of Napoli and its active volcanoes. Geological Field Trips and Maps. 2019;**11**(1):2-107

[15] Poret M, Finizola A, Ricci T, Ricciardi GP, Linde N, Mauri G, et al. The buried caldera boundary of the Vesuvius 1631 eruption revealed by

present-day soil CO2 concentration. Journal of Volcanology and Geothermal Research. 2019;**375**:43-56

[16] Porreca M, Pimentel A, Kueppers U, Izquierdo T, Pacheco J, Queiroz G. Event stratigraphy and emplacement mechanisms of the last major caldera eruption on Sete Cidades volcano (São Miguel, Azores): The 16 ka Santa Barbara formation. Bulletin of Volcanology. 2018;**80**(76). https://doi.org/10.1007/s00445-018-1250-0

[17] Sparice D, Scarpati C, Mazzeo FC, Petrosino P, Arienzo I, Gisbert G, et al. New proximal tephras at Somma-Vesuvius: Evidences of a pre-caldera, large (?) explosive eruption. Journal of Volcanology and Geothermal Research. 2017;**335**:71-81

[18] Miyabuchi Y, Okuno M, Torii M, Yoshimoto M, Kobayashi T. Tephrostratigraphy and eruptive history of post-caldera stage of Toya Volcano, Hokkaido, northern Japan. Journal of Volcanology and Geothermal Research. 2014;**281**:34-52

[19] Ventura G, Vilardo G, Bronzino G, Gabriele G, Nappi R, Terranova C. Geomorphological map of the Somma-Vesuvius volcanic complex (Italy). Journal of Maps. 2005;**30**:30-37

[20] Sigurdsson H, Carey S, Cornell W, et al. The eruption of Vesuvius in A.D. 79. National Geographic Research and Exploration. 1985;**1**:332-387